Learning and Analytics in Intelligent Systems

Volume 22

The main aim of the series is to make available a publication of books in hard copy form and soft copy form on all aspects of learning, analytics and advanced intelligent systems and related technologies. The mentioned disciplines are strongly related and complement one another significantly. Thus, the series encourages cross-fertilization highlighting research and knowledge of common interest. The series allows a unified/integrated approach to themes and topics in these scientific disciplines which will result in significant cross-fertilization and research dissemination. To maximize dissemination of research results and knowledge in these disciplines, the series publishes edited books, monographs, handbooks, textbooks and conference proceedings.

More information about this series at http://www.springer.com/series/16172

Maria Virvou · George A. Tsihrintzis ·
Lefteri H. Tsoukalas · Lakhmi C. Jain
Editors

Advances in Artificial Intelligence-based Technologies

Selected Papers in Honour of Professor Nikolaos G. Bourbakis—Vol. 1

Springer

Editors
Maria Virvou
Department of Informatics
University of Piraeus
Piraeus, Greece

George A. Tsihrintzis
Department of Informatics
University of Piraeus
Piraeus, Greece

Lefteri H. Tsoukalas
Director of Applied Intelligent Systems Laboratory
Department of Nuclear Engineering
Purdue University
West Lafayette, IN, USA

Lakhmi C. Jain
KES International
Shoreham-by-Sea, UK

University of Technology Sydney
Sydney, Australia

Liverpool Hope University
Liverpool, UK

ISSN 2662-3447 ISSN 2662-3455 (electronic)
Learning and Analytics in Intelligent Systems
ISBN 978-3-030-80570-8 ISBN 978-3-030-80571-5 (eBook)
https://doi.org/10.1007/978-3-030-80571-5

This Springer imprint is published by the registered company Springer Nature Switzerland AG
The registered company address is: Gewerbestrasse 11, 6330 Cham, Switzerland

Foreword

Artificial Intelligence (AI) is the science that studies the "evolution and behavior" of intelligent systems.

The central goal of Artificial Intelligence is not only to integrate intelligence into computers, but rather to turn computers into more useful machines. The scientific field of Artificial Intelligence is neither new nor easy to outline and define. It can be found in the connection of many concepts, such as perception, knowledge, logic, and processes (learning, decision, planning, etc.) in an attempt to understand the principles and theories that allow intelligence and their transformation into algorithmic processes. As an interdisciplinary academic field, it has appeared as early as 1950 [1,2]. Since then, its evolution has not been linear, as it has often relied on and expanded upon many and diverse other disciplines, including optimization theory and practice, decision-making, mathematical theories of fuzzy logic and statistics, neurobiology and paradigms of biological intelligence, psychology, cognition, pedagogical theories, and philosophy. In its course through the past six decades, Artificial Intelligence has met with success but also disappointment due to misconception of its complexity and the high expectations of users and industry [2–4]. While in many cases Artificial Intelligence achieves a performance significantly superior to that of humans, especially in the association of multiple data sets and mining, its advances still place technology very far away from building machines with the capacity to understand or learn any intellectual task that a human being can and even further away from building machines that can experience consciousness and emotionality [2].

Recently, the entire field of Artificial Intelligence is pressed toward further research and development following important and evolving new technological advancements in big data, mobile software applications and smartphones, Internet of Things, Computer Vision, business, manufacturing, and applications in essentially all sorts of human activities and professions. Because of the enormous pace of growth in new theories and application areas and high rates of success, *Artificial Intelligence-based Technologies* are already having a major impact on the society itself and affect positively our daily lives [3–6]. Undoubtedly, an impact of this magnitude also includes risks and threats in security, privacy, safety, transparency, business, competition, the job market, fundamental rights, democracy, or even human existence itself [7–10]. Artificial Intelligence-based Technologies need to be developed under careful human supervision and by highly qualified scientists who understand the associated risks and threats.

Professor **Nikolaos G. Bourbakis** certainly qualifies as a world-outstanding researcher in various fields of Artificial Intelligence which he has been serving from various posts for almost fifty years now. Holding a B.S. degree in Mathematics from the National and Kapodistrian University of Athens, Greece, a Certificate in Electrical Engineering from the University of Patras, Greece, and a Ph.D. degree in Computer Engineering and Informatics (awarded with excellence) from the Department of Computer Engineering & Informatics, University of Patras, Greece, he is currently a Distinguished Professor of Information & Technology and the Director of the Center of Assistive Research Technologies (CART) at Wright State University, Ohio, USA. For his many achievements in Artificial Intelligence-based Technologies, he has received many distinctions and awards, including elevation to IEEE Fellow (1996); IEEE Computer Society Technical Research Achievement Award; Member of the New York Academy of Sciences; Diploma of Honor in Artificial Intelligence, School of Engineering, University of Patras, Greece; ASC Outstanding Scientists & Engineers Research Award; Dr. F. Russ IEEE Biomedical Engineering Award, Dayton Ohio; Recognition Award for Outstanding Scholarly Achievements and Contributions in the field of Computer Science, University of Piraeus, Greece; IEEE EMBS-GR Award of Achievements; IEEE Computer Society 30 years ICTAI Outstanding Service & Leadership Recognition; Honorary Doctorate Degree of the University of Piraeus, Greece (2020).

The book at hand has been edited in Professor Nikolaos G. Bourbakis' honor and constitutes a streamline and coherent account of Advances in Artificial Intelligence-based Technologies. It consists of 12 chapters, which have been written by active and recognized researchers and report on recent research and development findings. Overall, the book is well structured as, besides an editorial/introductory chapter, it has been further divided into three parts devoted to *Advances in Artificial Intelligence Tools and Methodologies* (4 chapters), *Advances in Artificial Intelligence-based Applications and Services* (5 chapters), and *Theoretical Advances in Computation and System Modeling* (2 chapters).

The breadth and depth of the coverage of Artificial Intelligence-based Technologies is impressive. I am confident that expert readers in the various Artificial Intelligence and Computer Science disciplines will find it helpful in their researches. I am also confident that the book will as well appeal to general readers with an interest in finding out about current Artificial Intelligence-based Technologies.

I congratulate the editors on both undertaking the task of honoring an outstanding researcher, such as Nikolaos G. Bourbakis, and on presenting this timely book to the Artificial Intelligence and Computer Science communities and the general reader.

Prof. Dr. Euripides G. M. Petrakis
Director of the Intelligent Systems Laboratory
School of Electrical and Computer Engineering
Technical University of Crete
Crete, Greece

Further Reading

1. A. M. Turing, "Computing Machinery and Intelligence", *Mind* vol.49, pp. 433–460, 1950.
2. R. Anyoha, The History of Artificial Intelligence, Aug. 2017 https://sitn.hms.harvard.edu/flash/2017/history-artificial-intelligence/
3. From Industry 4.0 to Society 5.0: The Big Societal Transformation Plan of Japan. https://www.i-scoop.eu/industry-4-0/society-5-0/
4. Society 5.0 https://www8.cao.go.jp/cstp/english/society5_0/index.html
5. N. Bourbakis (ed.), *Artificial Intelligence Methods and Applications*, World Scientific Publishing Company, ISBN-13: 978-9810210571, Dec. 1992, 732 pages
6. N. Bourbakis, "Symbiotic and Intelligent Robotics", in A. Kent and J. G. Williams (Eds.), **Computer Science and Technology Encyclopedia**, Marcel Dekker Inc., 1991, vol. 25, pp. 332–378, (invited chapter)
7. S. Zuboff, *The Age of Surveillance Capitalism - The Fight for a Human Future at the New Frontier of Power*, PublicAffairs Pub., ISBN-101541758005, 2020, March 2020, 704 pages
8. N. Bostrom and M. M. Ćirković (eds.), *Global Catastrophic Risks*, Oxford University Press, ISBN-13: 978-0199606504, 2011
9. N. Bostrom, *Superintelligence: Paths, Dangers, Strategies*, Oxford University Press, ISBN-13: 978-0198739838, 2016
10. N. Bourbakis, "Artificial Intelligence (AI) and its Impact to Humanity: Immortality or Last Invention", Invited Keynote Lecture, University of Piraeus, Greece, Feb. 6, 2020

Preface

A world-recognized researcher can be honored in a variety of ways, including elevation of his professional status or various prestigious awards and distinctions. When, additionally, the same researcher has served as an advisor to generations of undergraduate, graduate, and doctoral students and as mentor to faculty and colleagues, the task of appropriately honoring him becomes even harder! **Perhaps, the best way to honor this person is to ask former doctoral students, as well as colleagues and fellow researchers from around the world, to include some of their recent research results in one or more high quality volumes edited in his honor. Such an edition indicates that other researchers are pursuing and extending further what they have learned from him in research areas where he made outstanding contributions**.

Professor **Nikolaos G. Bourbakis** has been serving the fields of Artificial Intelligence (including Machine Learning/Deep Learning) and Assistive Technologies from various posts for almost fifty years now. He received a BS in Mathematics from the National and Kapodistrian University of Athens, Greece, a Certificate in Electrical Engineering from the University of Patras, Greece, and a PhD in Computer Engineering and Informatics (awarded with excellence), from the Department of Computer Engineering & Informatics, University of Patras, Greece.

Dr. Bourbakis (IEEE Fellow-1996) is currently a Distinguished Professor of Information & Technology and the Director of the Center of Assistive Research Technologies (CART) at Wright State University, Ohio, USA. He is the founder and Editor-in-Chief of the International Journal on Artificial Intelligence Tools, the International Journal on Monitoring and Surveillance Technology Research (IGI-Global, Publ.), and the EAI Transactions on Bioengineering & Bioinformatics. He is also the Founder and Steering Committee Chair of several International IEEE Computer Society Conferences (namely, ICTAI, ICBIBE, ICIISA), Symposia, and Workshops. He pursues research in Assistive Technologies, Applied Artificial Intelligence, Bioengineering, Information Security, and Parallel/Distributed Processing, which is funded by USA and European government and industry. He has published extensively in IEEE and International Journals and he has graduated, as the main advisor, several dozens of doctoral students. His research work has been internationally recognized and he has received several prestigious awards, including: IEEE Computer Society Technical Research Achievement Award; Member of the New York Academy of Sciences; Diploma of Honor in AI School of Engineering, University of Patras, Greece; ASC Outstanding Scientists & Engineers Research Award; Dr. F. Russ IEEE Biomedical Engineering award, Dayton Ohio; Most Cited Article in Pattern Recognition Journal; IEEE ICTAI and ICBIBE best paper Awards; Recognition Award for Outstanding Scholarly Achievements and Contributions in the field of Computer Science, University of Piraeus, Greece; IEEE EMBS-GR Award of Achievements; IEEE Computer Society 30 years ICTAI Outstanding Service & Leadership Recognition; Honorary Doctorate degree of the University of Piraeus, Greece.

We have been collaborating with Professor Nikolaos G. Bourbakis for very many years. Thus, we proposed and undertook with pleasure the task of editing a special book in his honor. The response from former mentees, colleagues, and fellow researchers of his has been great! Unfortunately, page limitations have forced us to limit the works to be included in the book and we apologize to those authors whose works were not included. Despite the decision not to include all proposed chapters in the book, it became apparent that not only one, but **three** volumes of the special book had to be developed, each of which would focus on different aspects of Dr. Nikolaos G. Bourbakis's research activities.

The book at hand constitutes the first volume and is devoted to **Advances in Artificial Intelligence-based Technologies**. While honoring Professor Nikolaos G. Bourbakis, this book also serves the purpose of exposing its reader to some of the most significant advances in Artificial Intelligence-based Technologies. As such, the book is directed toward professors, researchers, scientists, engineers, and students in computer science-related disciplines. It is also directed toward readers who come from other disciplines and are interested in becoming versed in some of the most recent Artificial Intelligence-based Technologies. We hope that all of them will find it useful and inspiring in their works and researches.

We are grateful to the authors and reviewers for their excellent contributions and visionary ideas. We are also thankful to Springer for agreeing to publish this book in its **Learning and Analytics in Intelligent Systems** series. Last, but not least, we are grateful to the Springer staff for their excellent work in producing this book.

Piraeus, Greece — George A. Tsihrintzis
Piraeus, Greece — Maria Virvou
Lafayette, Indiana, USA — Lefteri H. Tsoukalas
Vietri, Italy — Anna Esposito
Sydney, Australia — Lakhmi C. Jain

Contents

Chapter 1
Introduction to Advances in Artificial Intelligence-Based Technologies

Maria Virvou, George A. Tsihrintzis, Lefteri H. Tsoukalas, and Lakhmi C. Jain

Abstract As human societal organization is being restructured into, so-called, "*Society 5.0*", the field of Artificial Intelligence-based technologies is growing continuously and rapidly and expands in itself while its results are applied to an ever-increasing number of other disciplines. The book at hand aims at exposing the reader to some of the most significant recent advances in Artificial Intelligence-based technologies, while at the same time honouring one of the most significant researchers in Applied Artificial Intelligence, namely Professor Nikolaos G. Bourbakis. As such, the book is directed towards professors, researchers, scientists, engineers and students in all computer science-related disciplines. It is also directed towards readers who come from other disciplines and are interested in becoming versed in some of the most recent Artificial Intelligence-based technologies. An extensive list of bibliographic references at the end of each chapter guides its readers to probe further into the application areas of interest to them.

Editorial Note

With the 4th Industrial Revolution ongoing [1, 2] and human societal organization being restructured into, so-called, "Society 5.0" [3, 4], the field of Artificial Intelligence-based technologies is growing continuously and rapidly and expands

M. Virvou · G. A. Tsihrintzis (✉)
Department of Informatics, University of Piraeus, 18534 Piraeus, Greece
e-mail: geoatsi@unipi.gr

M. Virvou
e-mail: mvirvou@unipi.gr

L. H. Tsoukalas
Department of Nuclear Engineering, Purdue University, West Lafayette, IN, USA
e-mail: tsoukala@purdue.edu

L. C. Jain
Liverpool Hope University, Liverpool, UK

University of Technology Sydney, Sydney, NSW, Australia

M. Virvou et al. (eds.), *Advances in Artificial Intelligence-based Technologies*, Learning and Analytics in Intelligent Systems 22,
https://doi.org/10.1007/978-3-030-80571-5_1

in itself, while its results are applied to an ever-increasing number of other disciplines. Researchers worldwide aim at incorporating cognitive abilities into machines, such as learning and problem solving. When machines and software systems have been enhanced with Artificial Intelligence components, they become better and more efficient at performing specific tasks.

The advances of Artificial Intelligence and its technological applications already affect many aspects of everyday life, the workplace and people's homes, as well as human relationships on the levels of both professional and social interactions. In the foreseeable and not-so-distant future, Artificial Intelligence is expected to affect yet more aspects of life. This is due to the fact that researchers are perpetually faced with new technological challenges which are often interrelated and, sometimes, involved in disciplines other than Artificial Intelligence or, more generally, computer science, in which Artificial Intelligence is expected to play an important role [5–7]. For example, the *Internet of Things* (IOT) constitutes one of the pillars of the 4th Industrial Revolution and is already "invading" everyday life and shaping Society 5.0 with new technologies, such as smart homes [8], smart cities [9, 10], smart industries [11] or smart healthcare [12, 13], all combined with the need and task to protect all kinds of data from malicious abuse [14–16].

Clearly, Artificial Intelligence stands out as a research discipline due to its worldwide pace of growth, corresponding success rates and impact on science, technology and society. More specifically, significant progress is made along the directions of (i) theoretical advances, (ii) innovative tools and methodologies and (iii) exploration of new areas of application.

The book at hand aims at exposing its readers to some of the most significant advances in Artificial Intelligence-based technologies, including:

I. *Advances in Artificial Intelligence Tools and Methodologies*,
II. *Advances in Artificial Intelligence-based Applications and Services*, and
III. *Theoretical Advances in Computation and System Modeling.*

At the same time, the book also aims at honouring Professor Nikolaos G. Bourbakis, an outstanding researcher and educator in Applied Artificial Intelligence, who has conducted leading research in this field for approximately fifty years and has inspired, advised and mentored tens of students, fellow researchers and colleagues.

More specifically, the book at hand consists of the current editorial chapter (Chapter 1) and an additional eleven (11) chapters. All chapters in the book were invited from authors who work in the corresponding chapter theme and are recognized for their significant research contributions. In more detail, the chapters in the book are organized into three parts, as follows.

The *first part* of the book consists of four chapters devoted to *Advances in Artificial Intelligence Tools and Methodologies.*

Specifically, Chapter 2, by K. Michalopoulos, G. Bebis, and N.G. Bourbakis, is entitled "*Synthesizing 2D Ground Images for Maps Creation and detecting Texture Patterns.*" The authors present a methodology for the generation of paths for surveillance ground maps, which is based on detecting and monitoring texture patterns across consecutive images. Applications of their work include autonomous path planning of aerial vehicles.

Chapter 3, by G. Gaudi, B. Kapralos K.C. Collins, and A. Quevedo, is entitled "*Affective Computing: An Introduction to the Detection, Measurement, and Current Applications*." The authors' aim is to provide the reader with a greater understanding of affective computing while highlighting current issues, example use cases, limitations, and areas of future research.

Chapter 4, by P. Krasadakis, E. Sakkopoulos and V.S. Verykios, is entitled "*A Database Reconstruction Approach for the Inverse Frequent Itemset Mining Problem*." The authors present a novel hiding scheme that produces a sanitized database based on the ideal borders that are computed through the application of a constraint-based frequent itemset mining algorithm.

Chapter 5, by Z. Kokosinski and K. Jaworski, is entitled "*A Rough Inference Software System for Computer-Assisted Reasoning*." The authors present an Artificial Intelligence Tool designed to assist in the building and processing of decision tables, the reduction of attributes and the generation of non-redundant rules that support the decision process in information systems.

The *second part* of the book consists of five chapters devoted to *Advances in Artificial Intelligence-based Applications and Services*.

Specifically, Chapter 6, by K. Dimitropoulos and I. Hatzilygeroudis, is entitled "*Context Representation and Reasoning in Robotics-An Overview*." The authors review basic research work focusing on representation the representation of contextual information with respect to robotics and the interaction between humans and robots and propose directions that future work may follow in this area.

Chapter 7, by A. Kontogianni, E. Alepis and C. Patsakis, is entitled "*Smart Tourism & Artificial Intelligence: Paving the Way to the post-COVID-19 Era*." The authors focus on Artificial Intelligence approaches presented by both researchers and practitioners in the tourism industry and analyse exploited technologies which show the potential to reboot the tourism industry and help it recover from the serious losses it suffered due to the COVID-19 pandemic.

Chapter 8, by M. Alamaniotis, is entitled "*Challenges and AI-based Solutions for Smart Energy Consumption in Smart Cities*." The author discusses challenges and Artificial Intelligence-driven technological solutions pertaining to smart energy services within the context of a smart city.

Chapter 9, by G. Molnár, is entitled "*How to Make Different Thinking Profiles Visible Through Technology: The Potential for Log File Analysis and Learning Analytics*." The author aims at showing how innovations in technology-based assessment and the options available via educational data mining and log file analysis can provide answers to research questions which remain unanswered with traditional assessment techniques and can quantify qualitative differences by making thinking visible.

Chapter 10, by D.C. Gkikas and P.K. Theodoridis, is entitled "*AI in Consumer Behavior*." The authors attempt to map the stages of Artificial Intelligence contribution to consumer behavior describing the essential marketing milestones.

The *third part* of the book consists of two chapters devoted to *Theoretical Advances in Computation and System Modeling.*

Specifically, Chapter 11, by M. Bonnin, F. L. Traversa and F. Bonani, is entitled "*Coupled Oscillator Networks for von Neumann and non-von Neumann computing.*" The authors provide a panorama of the most relevant alternative computing techniques, both according and not-according to the von Neumann architecture, highlighting which of the classical challenges, such as energy efficiency and/or computational complexity, the alternatives are trying to tackle.

Finally, Chapter 12, by R.-E. Tudoroiu, M. Zaheeruddin, S. M. Radu, D. D. Burdescu, and N. Tudoroiu, is entitled "*Design and Implementation in a New Approach of Non-minimal State Space Representation of a MIMO Model Predictive Control Strategy-Case Study and Performance Analysis.*" The authors investigate a new design approach of a model predictive control strategy in closed-loop applied to minimal and non-minimal state-space realizations implemented in real-time on a MATLAB SIMULINK platform and illustrate the effectiveness of the proposed control strategy on the case study of a centrifugal chiller system.

Book Summary And Future Volumes

In this book, we have presented some very significant advances in Artificial Intelligence-based technologies, while honouring Professor Nikolaos G. Bourbakis who has been conducting leading research in Applied Artificial intelligence for approximately fifty years and has inspired, advised and mentored tens of students, fellow researchers and colleagues. The book is directed towards professors, researchers, scientists, engineers and students in computer science-related disciplines. It is also directed towards readers who come from other disciplines and are interested in becoming versed in some of the most recent Artificial Intelligence-based technologies. We hope that all of them will find it useful and inspiring in their works and researches.

On the other hand, societal demand continues to pose challenging problems, which require ever more advanced theories and ever more efficient tools, methodologies, systems and Artificial Intelligence-based technologies to be devised to address them. Thus, the readers may expect that additional related volumes will appear in the future.

References

1. J. Toonders, Data is the new oil of the digital economy, *Wired*, https://www.wired.com/insights/2014/07/data-new-oil-digital-economy/
2. K. Schwabd, The fourth industrial revolution - what it means and how to respond, Foreign Affairs (2015), https://www.foreignaffairs.com/articles/2015-12-12/fourth-industrial-revolution
3. From Industry 4.0 to Society 5.0: the big societal transformation plan of Japan, https://www.i-scoop.eu/industry-4-0/society-5-0/
4. Society 5.0, https://www8.cao.go.jp/cstp/english/society5_0/index.html

5. G.A. Tsihrintzis, D.N. Sotiropoulos, L.C. Jain (eds.), *Machine Learning Paradigms – Advances in Data Analytics*, vol. 149 in Intelligent Systems Reference Library Book Series (Springer, 2018)
6. G.A. Tsihrintzis, M. Virvou, E. Sakkopoulos, L.C. Jain (eds.), *Machine Learning Paradigms - Applications of Learning and Analytics in Intelligent Systems*, vol. 1 in Learning and Analytics in Intelligent Systems Book Series (Springer, 2019)
7. M. Virvou, E. Alepis, G.A. Tsihrintzis, L.C. Jain, *Machine Learning Paradigms – Advances in Learning Analytics*, vol. 158 in Intelligent Systems Reference Library Book Series (Springer, 2020)
8. H. Jiang, C. Cai, X. Ma, Y. Yang, J. Liu, Smart home based on WiFi sensing: a survey. IEEE Access **6**, 13317–13325 (2018)
9. Du. Rong, P. Santi, M. Xiao, A.V. Vasilakos, C. Fischione, The sensable city: a survey on the deployment and management for smart city monitoring. IEEE Commun. Surv. Tutorials **21**(2), 1533–1560 (2019)
10. B.P.L. Lau, S.H. Marakkalage, Y. Zhou, N.U. Hassan, C. Yuen, M. Zhang, U.X. Tan, *A survey of data fusion in smart city applications*. Inf. Fusion **52**, 357–374 (2019)
11. X. Liu, J. Cao, Y. Yang, S. Jiang, CPS-based smart warehouse for industry 4.0: a survey of the underlying technologies. Computers **7**(1), 13 (2018)
12. M.M. Dhanvijay, S.C. Patil, Internet of Things: a survey of enabling technologies in healthcare and its applications. Comput. Networks **153**, 113–131 (2019)
13. Q. Cai, H. Wang, Z. Li, X. Liu, A survey on multimodal data-driven smart healthcare systems: approaches and applications. IEEE Access **7**, 133583–133599 (2019)
14. M. Khawla, M. Tomader, A survey on the security of smart homes: issues and solutions, in *ICSDE* 2018, pp. 81–87
15. D. Eckhoff, I. Wagner, Privacy in the smart city - applications, technologies, challenges, and solutions IEEE Commun. Surv. Tutorials **20**(1), 489–516 (2018)
16. A. Algarni, A survey and classification of security and privacy research in smart healthcare systems. IEEE Access **7**, 101879–101894 (2019)

Part I
Advances in Artificial Intelligence Tools and Methodologies

Chapter 2
Synthesizing 2D Ground Images for Maps Creation and Detecting Texture Patterns

K. Michalopoulos, G. Bebis, and N. Bourbakis

Abstract This chapter presents a methodology for the generation of paths from surveillance ground maps. The methodology detects and monitors texture patterns across consecutive images using an online sequential algorithm that creates piecewise linear approximation of the monitored trajectory. It takes under consideration the local structure of the patterns in order to extract a local path. The overall global path is sequentially formed from each segment. We present the preprocessing steps used to extract patterns of interest from an image and the results from two case study examples. The methodology has applications in the autonomous path planning of aerial vehicles following a specified structure, such as that of a river or road.

Keywords Texture patterns · Texture paths · Image analysis

2.1 Introduction

Image processing and computer vision have been widely used for the processing, analysis and information extraction from aerial or satellite images [1–4], just to mention a few. Most of the applications try to find certain structures or objects as buildings, roads and vehicles. The goal is to characterize and recognize the scene under consideration and then study the changes over time. This applies whether we are dealing with urban, rural or areas of natural significance.

A different kind of task, frequently encountered, is to find and track a certain object or feature through a series of images or video. This problem has many applications in surveillance and activity monitoring to analysis of aerial images [5–7]. The steps involved for this kind of task include registration of the different captured

K. Michalopoulos · N. Bourbakis (✉)
CART Center, Wright State University, Dayton, OH, USA
e-mail: nikolaos.bourbakis@wright.edu

G. Bebis
University of Nevada, Reno, NV, USA

M. Virvou et al. (eds.), *Advances in Artificial Intelligence-based Technologies*, Learning and Analytics in Intelligent Systems 22,
https://doi.org/10.1007/978-3-030-80571-5_2

images, detection of the feature of interest in each image/frame and then modeling and description of the path that the object followed through the composed scene.

This can be considered as the problem of modeling a path from a given set of points in the scene. The goal in this case is to characterize the path and trajectories formed in the scene from features extracted from each frame. Tracking the movement of persons passing in front a surveillance camera and using this information to define walking paths in the scene is an application example [7].

In addition, advances in the technology of small manned or unmanned aerial vehicles have increased the potential applications of such vehicles in monitoring and surveillance of the natural environment and changes and have make the acquisition of aerial images affordable for many applications. They can be used to obtain images of a certain area in short periods of time, while keeping the total cost low. They can be equipped with different sensors that would allow the monitoring of different aspects of the considered landscape, like detection of river flooding or depth of a river [3]. In addition, flying monitoring devices like drones, can provide very valuable information in areas of natural or artificial disasters [19]. In such applications the point of interest is the spatial changes of the landscape and how they evolve over time. Modeling the spatial structure of natural landscape in a meaningful way is a difficult task and needs detailed a priori knowledge about the structure under consideration. That is why the navigation flying autonomous vehicles that monitor certain structures of the landscape, like a river or a road based on visual feedback is a challenging problem. In this case the problem is that the vehicle has to make decisions based on the limited information provided by the cameras and previous knowledge. Especially in natural environments where the structure can be unpredictable a way of modeling and forming a navigational path based on the available information is critical [8].

In this chapter we consider a sparse set of points in the scene and we want to extract the paths and trajectories defined by these points. We propose a methodology that divides the collected visual features based on their dominant direction and forms a navigation path on the unknown space.

Additionally, if we can adequately describe the paths that are formed in the scene from different features, we can use this information to classify different activities in relation with the paths that they follow in the scene. Each segment of the formed path contains all the information about the contained points/features and can be used to form a formal language, which can effectively describe the underlying structure [16].

Here the main assumption is that the detected points form altogether a meaningful set, meaning that all points originate from the same object or the same class of objects and we expect that they describe a meaningful spatial structure.

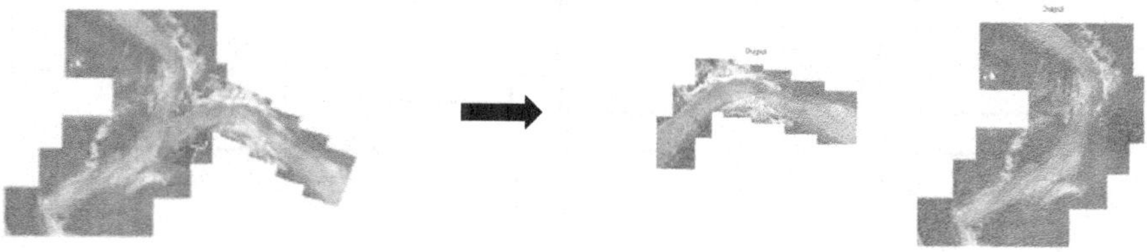

Fig. 2.1 Examples of the stiching process in sequential images for path-map generation

2.2 Synthesizing 2D Consecutive Region-Images for Space Map Generation

The use of a registration algorithm is essential when more than one images, captured from the same area, are used to compose the whole region of interest from which the paths need to be extracted. Applying the registration algorithm, we obtain the alignment of two consecutive images in order to achieve the stitching. Each new image is stitched with the previous output and so on until we receive the final image. For that application we used cross correlation to achieve the alignment of the two images. In addition, we assumed that the images are only translated from each other and we use the translation transformation in order to determine the parameters needed in registration procedure. One more robust way to register the images would be to use the affine transformation (scale and translation invariant) or SIFT features [12] (rotation invariant). The final output of registration of aerial images is depicted in Fig. 2.1. The same registration technique can be applied on underground tunnels as well. Then these ground maps are used for the detection of patterns in order to extract texture paths.

2.3 Texture Paths Detection

The proposed methodology in order to detect texture paths has to perform several synergistic steps introduced in [9] and significantly extended here. It starts with an original image, which is pre-partitioned into regions of interest and the removal of the background as a region of no interest. This procedure entails the image preprocessing step. The next step is to identify repeating texture blocks with specific features in the region of interest (foreground). After obtaining the desired patterns of texture blocks we go through the path detection algorithm, which is the final step in the proposed methodology. The whole procedure is illustrated in Fig. 2.2.

A. Image Preprocessing

Image preprocessing is considered as one of the most essential and critical steps in numerous computer vision and pattern recognition applications. The objective applying this very first step is to enhance the extraction of desirable features (in our

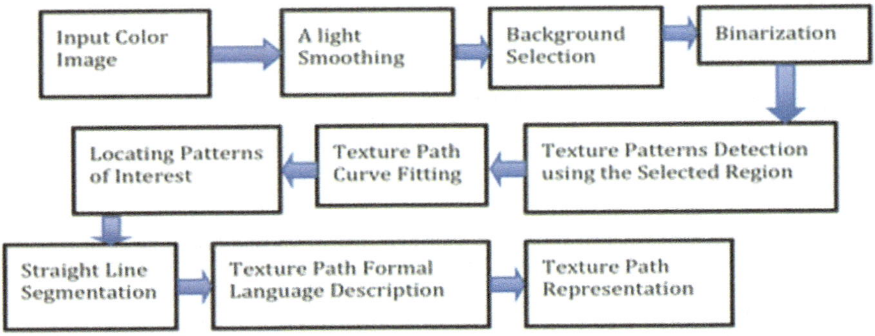

Fig. 2.2 The processing steps of the methodology

case texture blocks) while at the same time rejecting irrelevant information. Among the most well-known techniques used in image preprocessing are smoothing and noise elimination, filtering, edge sharpening, binarization, thinning and skeletonization, morphological operations such as erosion dilation and opening. The selection of the technique to be used its time is highly depend on the application and the nature of the problem. In our methodology we applied a light smoothing followed by binarization in order to achieve the background removal. By doing that the colors in this image were reduced to two, and additional artifacts that are irrelevant to the texture paths detection were eliminated.

Thus, the image-preprocessing step is essential for extracting the regions of interest from the captured images, under the assumption that we receive a sequence of images. The first task required is the elimination of the background which is accomplished by applying a smoothing in the image using bilateral filtering [1] followed by the Otsu's segmentation algorithm [2], which perform image thresholding. If we assume that the image is consisted of two classes of pixels the output of the Otsu's algorithm is a binary counterpart of the smoothed image. Then the foreground is selected using certain criteria of interest (texture, color, shape) and a simple region growing technique is applied to the foreground to extract the region of interest (ROI) [17]. The background represented with black pixels and the region of interest with white pixels. For instance, if we have a picture as is depicted in Fig. 2.3, then we perform image processing on it and the results are shown the Figs. 2.4, 2.5, 2.6, 2.7 below.

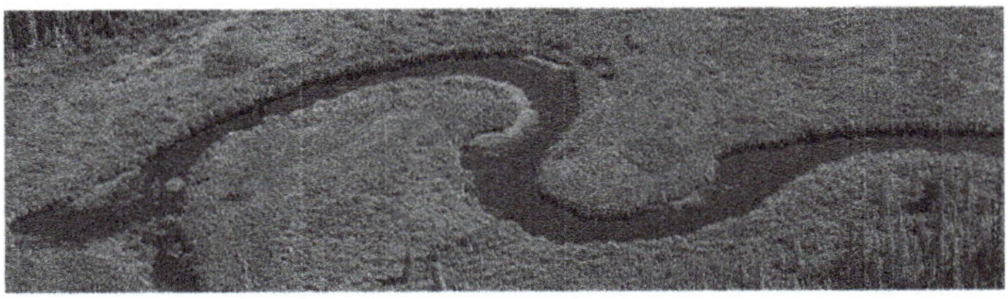

Fig. 2.3 Final grayscale image

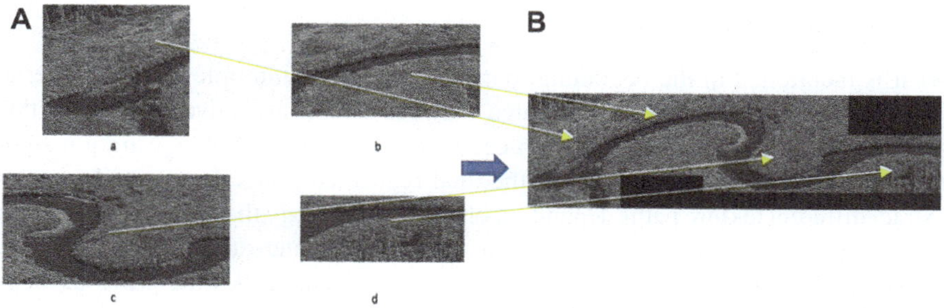

Fig. 2.4 **a** Consecutive image-regions of a scene. **b**. Full registered image

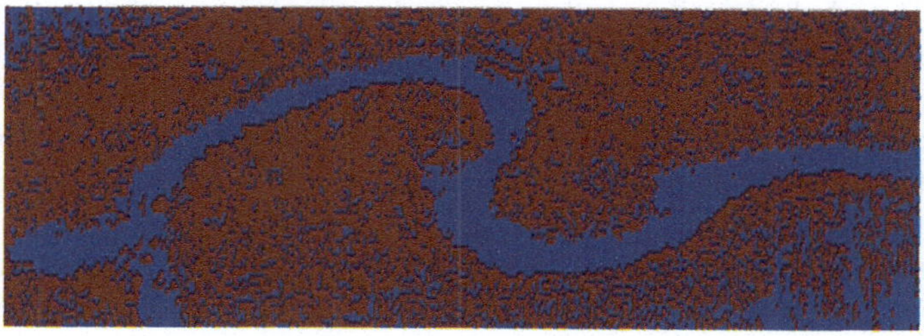

Fig. 2.5 Segmented image

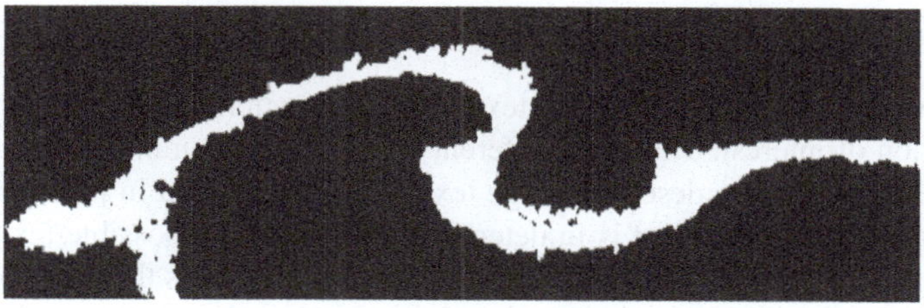

Fig. 2.6 Binary image

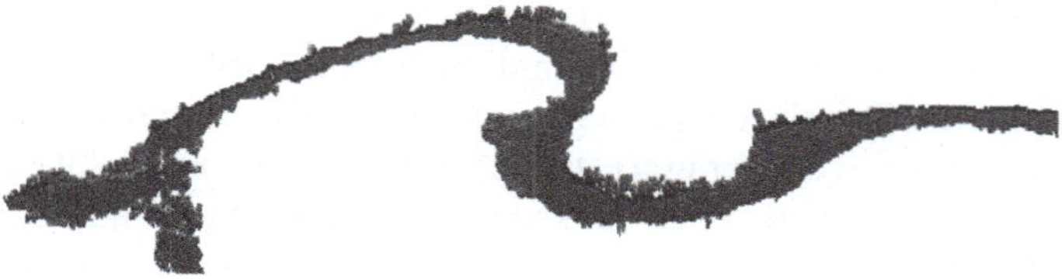

Fig. 2.7 Grayscale image of region of interest

Image Registration

As it is mentioned in the beginning, our input would be multiple sequential aerial images with a certain overlapping area between two consecutive images. In this scenario we will need to apply the procedure to all the images. A registration algorithm is essential to able to achieve the final registered image from which we can extract different texture paths depending on the methodology that is used. Figure 2.4a shows 4 consecutive images of a particular area. We are interested in finding specific patterns, which lie in the region of interest for each image. All the previous steps are applied for the first received image. The next step is the stitching of the first image with the second image in the queue using the cross-correlation algorithm. The same procedure is then repeated for the output of registration (registered two images). The whole procedure is then repeated until we receive the final image. For that application we assumed that the images are only translated from each other and we use the translation transformation in order to determine the parameters needed in registration procedure. One more robust way to register the images would be to use the affine transformation (scale and translation invariant) or SIFT features [5] (rotation invariant).

The final output (registered image) is depicted in Fig. 2.4b. Each intermediate output goes through the pattern extraction procedure to extract the texture paths. It is progressive procedure, which stops when we receive the final image of the scene.

The segmented image, the binary image and the grayscale image for the foreground are presented in Figs. 2.4, 2.5 and 2.6 respectively.

B. Texture Block-Patterns Detection

The next step is to identify repeating texture blocks (patterns) with specific features in the region of interest. There are different methods (structural, statistical, model-based and transform) to describe image texture as emphasized in [3, 4, 10, 11]. In the present work, our interest is in detecting and describing possible texture paths and representing them by a formal language approach towards this direction. Thus, we consider 3 × 3 square texture blocks with a predefined grayscale average value. We search the foreground for 3 × 3 squares that satisfy the criterion ($I_{avg} = T_{avg} \pm e$), where T_{avg} is the desired value and I_{avg} is the actual average value for a certain neighborhood in the image. Figure 2.6a shows two block-patterns that we plan to use in our work here. Since two block-patterns cannot form a path we consider textures that can have more than three repeating block-patterns. If a texture block-pattern is repeated multiple times and forms a path is considered as primitive. All the pixels in all the matching block-patterns are marked and are illustrated in Fig. 2.8.

Two different types of patterns were taken under consideration. In certain cases, the number of patterns in the image were not sufficient to demonstrate our methodology. To address this problem and be able to test the proposed formal language-based methodology for texture paths we decided to insert manually patterns in the ROI. The first step of the method is to identify those patterns and extract the centers that could be used as an input to the next stage. The results obtained, verify the accuracy of the algorithm used to extract the centers of the physically obtained patterns. Multiple

Fig. 2.8. 3 × 3 block-patterns in the ROI

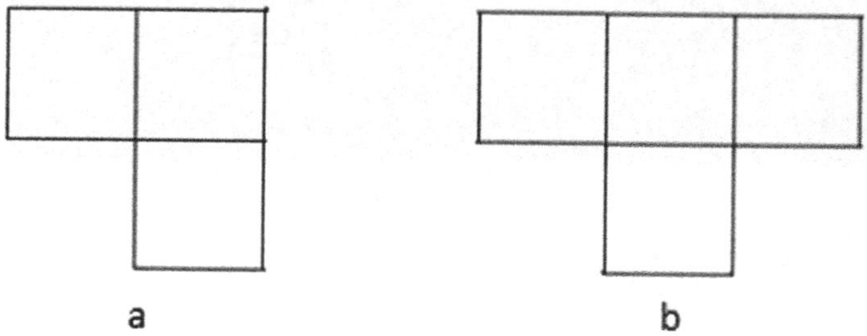

Fig. 2.9 The two patterns of interest used in the case study example

paths can be achieved, either from the different texture patterns or from the same texture pattern. For the example under consideration the selected texture patterns 1 and 2 are depicted in Fig. 2.9a and b.

The results from the patterns detection in the Fig. 2.3 are depicted in Fig. 2.10.

C. Path Detection Algorithm

In this effort here we implement an approach to path description that can be used in conjunction with a formal language as described in [9, 15]. The goal is to approximate the path in a linear way using a simple set of letters. In the proposed setup we use a rectangular in order to encode information about the whole area covered instead of just the path segment direction. So each rectangular represents a segment of the path and additional holds the number of points included in its area, the standard deviation of these point from the path, and the maximum distance from that path. Also each path segment encodes the distribution of points that lie above and below the line.

In order to determine the letters that will describe the path formed by the texture points we need to process the points and determine features that will allows us to describe the path following certain rules. We try to approximate the path by a set of points using straight-line segments. These segments describe the dominant linear direction of the points in a certain region of the scene. In this sense a single graph or a simple linear interpolation are not capable to describe the general trend of the points in the scene. On top of that, we need a compact description of the path that will allow us to approximate more features of the path other than direction and length.

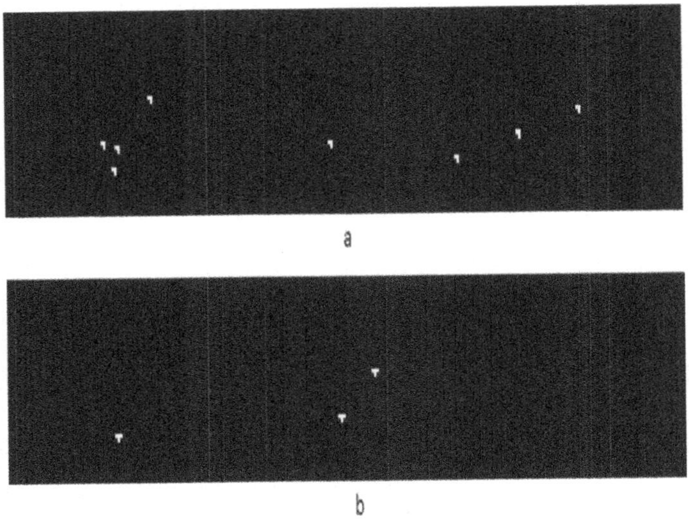

Fig. 2.10 **a** Total patterns 1 in the original image, **b** Total patterns 2 in the original image

Moreover, we want to identify junctions and other points of interest in the path. For these reasons, each path segment is defined as a path direction line inside a rectangle area covering all the local features/points and associated to a letter of the language. This formulation has allowed us, to define a formal language, capable to provide a meaningful representation of the paths that are formed from a set of points in terms of direction and compactness of the points [9, 16].

Provided that we have an initial point, the algorithm is capable to extract straight-line segments that represent the dominant direction of a set of local points (Fig. 2.11).

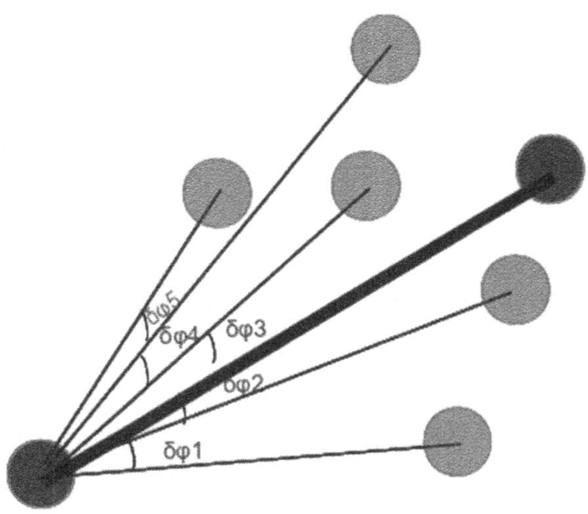

Fig. 2.11 Graphical illustration of the end point selection process. Each point in a certain neighborhood is used to create a line with the starting one. The difference of the angles formed by this line and the lines of the points that lie between the two points are calculated and noted in the Figure as $\delta\varphi_n$. The point that minimizes the weighted sum of the difference of angles is marked as end point

In essence we have a sequential clustering algorithm, which divides space into rectangular regions, using the dominant direction as feature. Each rectangular describes different conditions of the line segments that form the path. Each letter has information about the starting point the size of the rectangle, the direction and the distribution of the points inside the rectangle. All the points that their distance is less than the distance of the start and end point and lie inside a candidate rectangle will be called member points from now on for simplicity.

We used the dominant direction of the points under consideration as the criterion for determining the path direction (line inside the rectangle) and subsequently the area covered by the rectangle. For the calculation of the dominant direction we chose the line so that the sum of distances/angles of the points that lie between its two ends is minimum. We treat this as an error measure that we try to minimize. We normalize this measure by dividing with the number of member points for each candidate point. As the distance of the candidate point increases the number of member points increases and so does the sum of distances. It is straightforward that distant points will present higher error even though it would be desirable to include them in the same path segment. In order to avoid this behavior, we divide by the number of member points to balance this effect and allow the algorithm to select points further away.

The error measure can be further weighted so that the influence of local points to be greater than that of points further away. This step is necessary in case we are examining a large neighborhood with many points in order to preserve local information. We don't want a point in a path segment further away to mislead the algorithm to disregard paths defined by nearby points. Of course, the weighting function can be altered in order to suit the needs of the application and allow for different sensitivities. The function we seek to minimize is:

$$m(p_k) = \sum_{p_i} w(p_k) * |dist(p_k) - dist(p_i)|$$

We chose the weight function **w** to be proportional to the distance of the point from the origin and inversely proportional to the number of member points, in order to balance the effect of distance and number of points covered.

In order to define the direction that most of the points are pointing to, we take under consideration the lines from the starting initial point to all the other points in a certain region. Figure 2.11 displays the concept graphically. The steps in order to calculate the main features of each segment are described next.

We move the center of the axes to the initial point, in order to remove angle ambiguities and then we calculate the angles formed with the x-axis for all lines. Now instead from a set of coordinate points we have a set of angles between these points. Next we sort the points in terms of their distance from the initial point. For each line, we calculate the maximum distance of the points that lie between the starting and the current point. We also calculate the number of points that lie above or below the line.

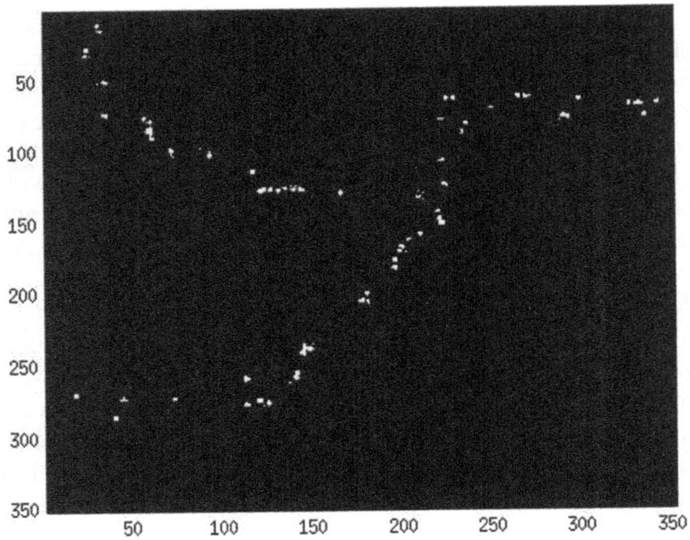

Fig. 2.12 It graphically shows the directions of two joined paths by using the specific patterns

For each point we calculate the difference of its angle to the angles formed by its member points. The point that presents the minimum value in the measure function is selected as the end point of the letter. The maximum distance from the line is used to define the sides of the rectangular and the distribution for the points the final letter to be used.

D. Junction Detection

An important feature of the algorithm is that it is able to detect junctions and classify paths that split somewhere in the scene into primary and secondary paths. The idea is based again in the notion of dominant direction only this time we use a weighted histogram in order to detect the dominant directions (Fig. 2.12).

In this step there is no need to define a letter, we only need to test whether the points ahead split and create a different distinct path. We count the peaks of the angles histogram and If the histogram is not flat enough mark the start point as junction. For example if we have two peaks that represent angles with 45° angle we denote the point as junction. The threshold of 45° is set arbitrarily and can be adjusted depending on the application. The end point that the initial algorithm found is used as the primary path, while for the secondary path the same procedure is applied using only the points in the opposite direction. Figure 2.13 graphically and progressively presents the angle histogram for a junction and the histogram of a single direction.

Curve Fitting

A well-known technique for approximating a line from a set of points is by finding the parameters of a function that would describe better the set of points in a least squares sense. The goal is to minimize the least squares of the discrete points and

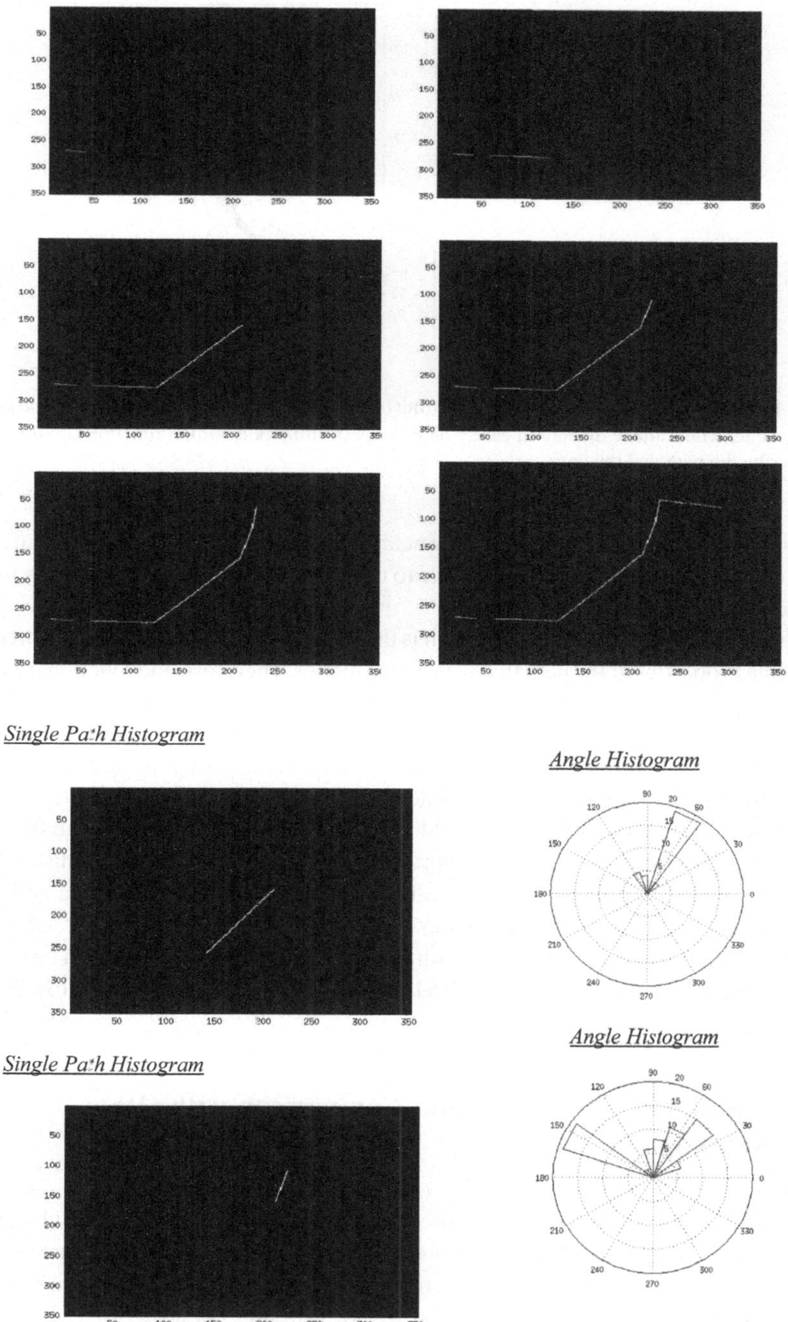

Fig. 2.13 It graphically shows the directions of a path and the angle of histogram

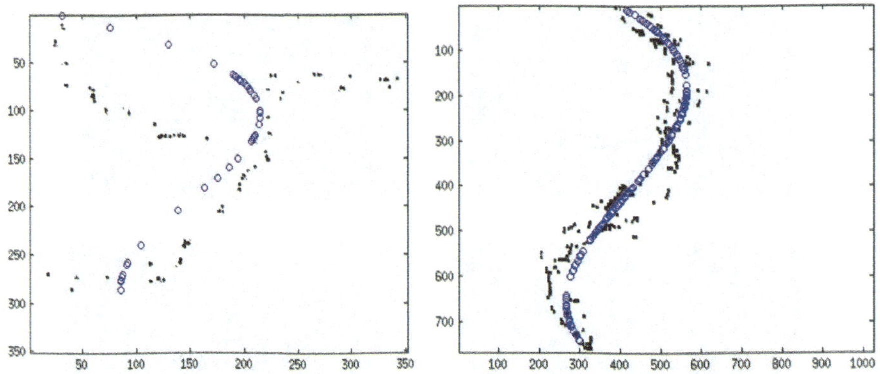

Fig. 2.14 Left image displays the resulting fitting of a third order polynomial function to the points in the scene. Right image displays a case where the best fitting polynomial function is not enough to describe the path and the data

the predicted points of the function. We can see in Fig. 2.14 an example of fitting a third order polynomial function in order to describe the path formed by points in the image.

The main problem with this approach is that usually we don't have any information about the properties of the path that we want to model. Therefore, the function used for approximation of the points is somewhat arbitrary. Therefore, the quality of the result depends on the function used for approximating the path. A significant limitation of this approach is that it cannot handle complex paths with junctions and the user has to carefully select the points or the scene to use for the curve fitting [18].

In order to limit this effect we could split the data into regions that can be fitted with the appropriate function. The approximation that we will get, if we use all the available points, will be different from the one if we use part of the data and then combine the partial results. In the latter case we would have to manually select the points to use in order to get a good result. Of course this approach does not eliminate the previous shortcomings of the method that still apply in each sub region examined.

2.4 Simulated Case Study and Comparison with Other Methods

The methodology described in the previous section was applied in aerial images and some of the results achieved are presented in the following subsections [9]. We assume that the texture blocks that we are interested in, are 3×3 blocks with a certain grayscale value and that the patterns that we are searching for are given in Fig. 2.15. As it is mentioned in the beginning, our input it may be multiple sequential aerial images with a certain overlapping area between two consecutive frames. To process the whole region of interest we need to apply a registration and a stitching algorithm.

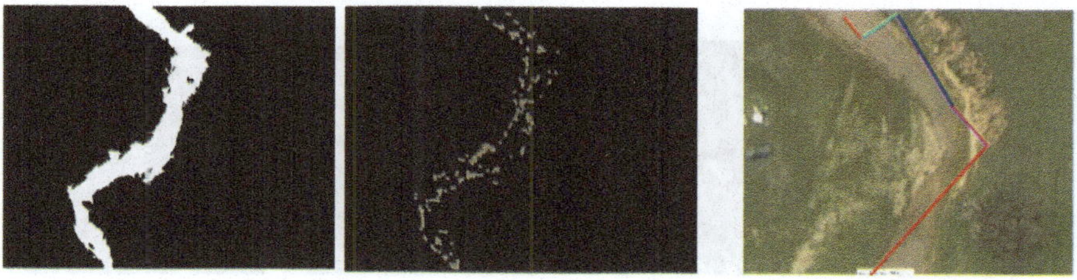

Fig. 2.15 First row displays the binarized image and the extracted texture patterns that form the path. Finally, the result of the algorithm for the a part of the image is presented in collumn 3. We can see that the extracted linear path follows the overal curve of the patterns and at the same time captures the local spatial variance of the patterns. The sensitivity of the algorithm in such variantions can be adjusted depending on the application

We apply the aforementioned methodology progressively to all of the intermediate outputs of the registration. Next, we will describe the pre-processing steps applied in the images and then the results of the algorithm are presented.

The first task required is the elimination of the background which accomplished by applying a smoothing in the image using bilateral filtering [13] followed by the Otsu's segmentation algorithm (Otsu Method [14]), which performs image thresholding. If we assume that the image is consisted of two classes of pixels the output of the Otsu's algorithm is a binary counterpart of the smoothed image. The selection of the foreground is performed by the user and then a simple region growing technique is applied to the foreground to extract the region of interest. The background represented with black pixels and the region of interest with white pixels. We will show the procedure for one of the images. The binary image is depicted in Fig. 2.15 (column 1). The next step was to extract the 3×3 blocks of that lie inside the foreground and present a specific grayscale value.

Next, we identified these patterns and extracted the centers that could be used as an input to the next stage. Multiple paths can be achieved, either from the different texture patterns or from the same texture pattern. The centers of detected patterns were used as input in for the path detection algorithm. The path detected for the proposed algorithm for the first 3 consecutive images is presented in Fig. 2.15 (row 1).

An interesting feature of the algorithm is that it is able to detect junctions and classify paths that split. The idea is based again in the notion of dominant direction only this time we use a weighted histogram in order to detect the dominant directions.

Here a path is defined as the curve that connects as many texture blocks as possible such that it has the minimum distance from all of them. Thus, two extreme solutions can be the paths produced by applying a graph traversing approach, like Prim's or Dikstra's algorithm and a curve fitting approach. The first approach takes into account the local distances of the nodes and creates path(s) that pass over the selected nodes. The second approach creates a path that considers the locations of the nodes globally and calculates the parameters of the polynomial curve that minimizes the distances of all nodes from it.

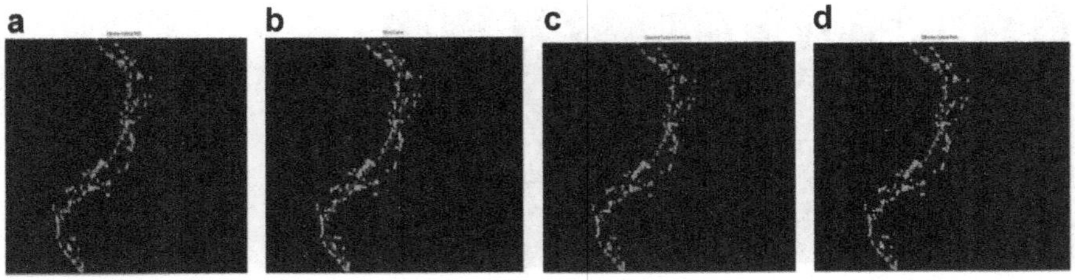

Fig. 2.16 It shows the results from four different approaches for path finding based on certain selected patterns. **a** Dijstra's **b** Fitted Curve, **c** Selected of texture centriods, and **d** Dijstra optimal path

To acquire a path using a graph traversing approach, we first create a fully connected graph of all the nodes and prune the longest edges, since it is more reasonable that the path is delineated by consequent nodes not far away. The longest distances are estimated adaptively, as those distances that are greater than the mean distance minus one standard deviation of the distance distribution. Then, the path that aims in satisfying the first constraint can be calculated using the Dijkstra's algorithm, which estimates the shortest path between two extreme nodes (closest to the frame's borders). We should note that a minimal spanning tree would satisfy completely this constraint, since it connects all the nodes of the graph, but that would not result in a smooth and continuous path.

The second constraint of the path finding problem is the contribution of all the nodes to the path's formation since there is no specific distinction among them. An obvious solution that satisfies this constraint is the polynomial curve that best fits the node points. Experimental results show that choosing a degree value of 3–6 for the polynomial offers a good trade-off between accuracy and complexity.

However, a solution that can jointly satisfy both constraints is expected to lead to a more natural path, that is a path that follows the general point distribution and at the same time traverses the nodes in small steps. Combination of the previous two approaches can yield such a solution. In this case, the polynomial curve is first fitted and is used as a guideline to refine the nodes, so as to select those that represent better the overall distribution. Specifically, only the nodes that do not exceed the median distance plus one standard deviation of the distance distribution from the curve are selected. The graph that connects them is then constructed and pruned same as before. Finding the optimal path on this reduced set of points using the Dijkstra's algorithm can now lead to the formation of an intuitively better path, as seen in Fig. 2.16.

2.5 Discussion

In this chapter we present a methodology for the identification and characterization of paths in aerial images of underground tunnels. The methodology can be used to trace the spatial structure formed by different patterns of interest in the scene.

The methodology outputs straight line approximations of the path without any prior knowledge about the distribution of the patterns in the scene. On top of that, each segment of the path holds information about the member patterns of the specific segment. Therefore, the local information of the path can be used to extract information from the specific segment. We also use a formal language based on the characteristics of different segments extracted by the proposed methodology in order to create a syntactic description of the obtained structure that can be used for learning and recognition.

Acknowledgements The authors thanks some programming contributions made by former students for implementing some parts of this project sponsored by CART Center.

References

1. S.S. Wegener, V.G. Ambrosia, J. Stoneburner, D.V. Sullivan, J.A. Brass, S.W. Buechel, R.G. Higgins, E.A. Hildum, S.M. Schoenung, D.L. Peterson, Demonstrating Acquisition of Real-Time Thermal Data over Fires Utilizing UAVs, (2002)
2. M.C. Quilter, V.J. Anderson, Low altitude/large scale aerial photographs: a tool for range and resource managers. Rangelands Arch. **22**(2), 13–17 (2006)
3. J. Lejot, C. Delacourt, H. Piégay, T. Fournier, M.-L. Trémélo, P. Allemand, Very high spatial resolution imagery for channel bathymetry and topography from an unmanned mapping controlled platform. Earth Surf. Process. Landforms **32**(11), 1705–1725 (2007)
4. L.L. Hess, J.M. Melack, E.M.L. Novo, C.C. Barbosa, M. Gastil, Dual-season mapping of wetland inundation and vegetation for the central Amazon basin. Remote Sens. Environ. **87**(4), 404–428 (2003)
5. I.N. Junejo, H. Foroosh, Euclidean path modeling from ground and aerial views, in *IEEE Conference on Computer Vision and Pattern Recognition, 2007. CVPR*, pp. 1–6 (2007)
6. I.N. Junejo, O. Javed, M. Shah, Multi feature path modeling for video surveillance, in *Proceedings of the 17th International Conference on Pattern Recognition, ICPR 2004*, vol. 2 (2004), pp. 716–719
7. D. Makris, T. Ellis, Path detection in video surveillance. Image Vis. Comput. **20**(12), 895–903 (2002)
8. S. Rathinam, P. Almeida, Z. Kim, S. Jackson, A. Tinka, W. Grossman, R. Sengupta, Autonomous searching and tracking of a river using an UAV, in *American Control Conference, 2007. ACC '07*, pp. 359–364 (2007)
9. N. Bourbakis, R. Patil, A methodology for automatically detecting texture paths and patterns in images, in *19th IEEE International Conference on Tools with Artificial Intelligence, 2007. ICTAI 2007*, vol. 1 (2007), pp. 504–512
10. A. Materka, M. Strzelecki, Texture analysis methods – a review, 1–33 (1998)
11. R.M. Haralick, K. Shanmugam, I. Dinstein, Textural features for image classification. IEEE Trans. Syst. Man Cybern. **3**(6), 610–621 (1973)
12. D.G. Lowe, Distinctive image features from scale-invariant keypoints. Int. J. Comput. Vision **60**(2), 91–110 (2004)
13. C. Tomasi, R. Manduchi, Bilateral filtering for gray and color images, in *Sixth International Conference on Computer Vision (IEEE Cat. No. 98CH36271)* (1998), pp. 839–846
14. P. Smith, D.B. Reid, C. Environment, L. Palo, P. Alto, P.L. Smith, A threshold selection method from gray-level histograms. **20**(1), 62–66 (1979)
15. N. Bourbakis, A heuristic collision free path planning for an autonomous platform. Int. J. Intell. Robot. Syst., Kluwer-Academic, **1**, 375–387 (1989)

16. H. Vafaie, N. Bourbakis, A tree grammar scheme for generation and recognition of simple texture paths in images, Proceedings IEEE International Symposium on Intelligent Control, VA (1988), pp. 201–206
17. A. Mogaddamzadeh, N. Bourbakis, A fuzzy technique for segmentation of colored images, in IEEE World Congress on Computational Intelligence, Fuzzy Conf. Orlando, FL (1994), pp. 83–88
18. K. Brown, N. Bourbakiss, Curve and surface fitting techniques in computer vision, Int. J. Image Graphics, 2021. 2150041 ISSN: 0219-4678
19. N. Bourbakis, I. Papadakis Ktistakis, T. Seleem, Smart cities detecting humans in regions of disasters: synergy of drones, micro-robots in underground tunnels. Int. J. Artif. Intell. Tools **29**(5), 2050006 (2020)

Chapter 3
Affective Computing: An Introduction to the Detection, Measurement, and Current Applications

Geoffrey Gaudi, Bill Kapralos, K. C. Collins, and Alvaro Quevedo

Abstract Affective computing aims to design and develop natural human-user interfaces that respond to the emotional needs of the user, bridging the gap between humans and technology. With the continuing technological advancements affective computing technologies are now available at the consumer level and are revolutionizing the ways in which we interact with computers. From simple entertainment applications to assistive technologies, the field of affective computing holds great promise. The aim of this chapter is to provide the reader with a greater understanding of affective computing while highlighting current issues, example use cases, limitations, and areas of future research.

Keywords Affective computing · Emotion · Artificial intelligence · Immersive technology

3.1 Introduction

In a 1950 paper titled "Computing Machinery and Intelligence", Alan Turing proposed a test designed to set a benchmark in computer intelligence: the ability of a human to determine if they were talking to another human or a machine [1]. Turing hinted at what would become a key differentiator in the ability of machines to demonstrate intelligence: emotion, and empathy. Turing recognized that while it would be relatively easy to develop a machine that could appear to answer logical or factual questions in a somewhat human manner (well developed in areas of natural language processing), to create a machine that not only appears to have empathy,

G. Gaudi · B. Kapralos (✉) · A. Quevedo
Software and Informatics Research Centre, Ontario Tech University, Oshawa, ON, Canada
e-mail: Bill.Kapralos@ontariotechu.ca

B. Kapralos
maxSIMhealth Group, Ontario Tech University, Oshawa, ON, Canada

K. C. Collins
School of Information Technology, Carleton University, Ottawa, ON, Canada

M. Virvou et al. (eds.), *Advances in Artificial Intelligence-based Technologies*,
Learning and Analytics in Intelligent Systems 22,
https://doi.org/10.1007/978-3-030-80571-5_3

but can process human emotion and respond in an appropriate manner would be a much more complex task. This latter portion of artificial intelligence (AI)—that is, the ability to detect, process and respond to emotional data—has become a defining goal of the field of affective computing.

Affective computing means a computer may be able to detect our increasing level of frustration and dispatch some assistance before we become overwhelmed. It may turn on the music we love as we walk into our house after a long day, knowing what music will uplift us, or calm us down, or put us in the right mood to hit the gym. It may provide a therapeutic ear to listen to our problems and suggest some other ways of looking at them. It may provide a friendly face and warm gesture as it explains to our grandfather one more time how to send us a video message. It may increase the intensity of the game we're playing as it watches us become bored, or flag our doctor that we're pushing the boundaries of our stress capacity and require assistance. It will tell companies how we feel about their products, or politicians how we feel about them or their political party, probably before we really understand such feelings ourselves.

Affective computing represents an enormous new set of tools that will revolutionize the ways in which we interact with computers (and other devices) in our lives. With the addition of new means of artificial intelligence handling the data processing and responses, affective computing promises to be one of the major advances of this decade.

The aim of this chapter is to provide the reader with a better understanding of affective computing while highlighting current issues, example use cases, limitations, and areas of future research. It is not designed to be a comprehensive overview of all tools and techniques available, but, rather, will highlight a few tools and explore how they are being used today.

3.2 Background

Affect is the expression of emotions or feelings, and affective computing refers to techniques for the detection, recognition, and prediction of human emotions with the goal of adapting computational systems to emotional states [2, 3]. Affective computing is advancing as a field that allows for new ways of human computer interaction [4]. Several fields and disciplines, such as virtual reality, video games and health monitoring already incorporate some of the tools and techniques of affective computing [5]. The role of emotion in both human cognition and perception, indicate that affective computers should not only provide better performance in assisting humans, but also enhance a computer's abilities to make decisions [6].

Our emotions play an important role in every facet of human interaction. They are involved in social interactions, perception, memory, learning, intelligence, and creativity [7], and are vital for us to function as rational decision-making human beings [6]. Emotions have evolved from simple mechanisms to ensure harm avoidance and attainment of vital resources into more complex ones that guide our social behaviours, and as such are closely intertwined with cognition [8]. Complex human behaviour emerges from dynamic interactions between multiple processes and brain networks. How we perceive our world, and the decisions we make, are determined by emotions. As such, any effective computer model of intelligence must take into account this complex relationship at play in our brains, and take into account emotional response.

However, emotions are not just cognitive neural processes: They also involve the body, and our psychophysiological responses are just that—a combination of our psychological response and our physiological processes. In the nineteenth century, William James and Carl Lange [9] argued that emotions result from physiological responses to external stimuli and that visceral and muscular bodily changes result from responses of the nervous system to stimuli. The awareness of those changes constitutes an emotion. For example, if an individual is in the presence of a snake, they are not afraid of the snake. The individual is aware of the bodily changes that the presence of the snake has triggered. They are afraid because they tremble. Physical changes, and emotions, are triggered immediately upon contact with the stimulus before cognitive perception occurs. At a very basic level, different core emotions can be distinguished from each other because each involves different bodily processes and changes [10]. Depending on the emotional response we experience, there may be changes in blood pressure, heart rate, thinking, and facial expressions, for example. Most emotional responses are fleeting, and we may not necessarily even be aware of them. Although emotional responses may vary, they have two common features: Firstly, emotions generate a global evaluation or core effect, that something is good or bad, helpful or harmful, approachable or to be avoided; and secondly, emotions generate some level of activation, or energy within us [11].

Appraisal theorists suggest that the relationship between emotions, personal goals and concerns are connected. This relationship is often suggested to be the basis for emotional elicitation and differentiation. For example, fear can be attached to a threatening stimulus, which may be life threatening. Our appraisal of a situation or event (e.g., safe, threatening, sad) leads to the reaction based on that appraisal. In appraisal theory, it is believed that emotions do not happen without physiological arousal. This theory, begun in the 1940s, is still being evolved and adapted today (cf. [8]), and other theories of emotion have also found favour, from cognitive-based theories (Schachter–Singer theory, Lazarus's Cognitive–mediational theory), neurological theories (e.g., [12]) and physiological theories (e.g., James–Lange theory, Cannon–Bard theory [9]), but all are still under dispute and no single dominant theory presides over the field. For an overview of the theories of emotion, see [13].

The Ortony, Clore, and Collins (OCC) model of emotion [14], relates the strength of a given emotion to events, agents, or objects in the environment. Six questions regarding the construction of emotions are posed. More specifically, if one were to build a surrogate person or robot, in order for it to be a believable agent, it must (i) share our sensory inputs, (ii) be equipped with more than eyes to see, (iii) ears that hear, (iv) muscles that move, (v) it must have affective reactions to be able to select what it looks at, and listens to, and (vi) select how it moves forward [14]. In living beings, these reactions occur at multiple levels from the neurochemical to the behavioural. According to the OCC model, everyday concepts of specific emotions are captured in prototypes and schemas, which play a formative role in understanding and recounting everyday experiences. More specifically, we infuse dramatic events with meaning, we reorganize, distil, summarize, and dramatize events into emotional events [14].

Many theories of emotion have attempted to categorise or classify emotions, leading to, for instance, the Geneva Emotion Wheel [15] and a similar approach, the Plotkin Emotion Wheel [16], and the two-dimensional valence-arousal model, commonly used in psychophysiological and some affective computing research (e.g., [17]). These two-dimensional models of emotion have, however, been criticized as insufficient to capture the nuances of human emotion, and it has been suggested that a four-dimensional approach is more appropriate to classifying emotion [18].

Emotion and affect are much disputed in terms of where and how in the body and brain they arise. Moreover, the difficulties with trying to articulate and classify emotion has been studied and argued for at least a century and scientific literature hints at the complexities of the field—not only are we trying to define what, exactly, we are measuring and responding to, but we are trying to teach machines something of which we don't yet have a complete understanding of ourselves. As such, it must be considered that affective computing is a field very much in its infancy, in a state of constant flux, and the tools and devices we use for detecting and measuring emotion (see below) are limited by this nascency.

3.3 Detection and Measurement Devices for Affective Computing

As we discussed above, emotion and affect are complex human processes involving cognitive and physical, neurobiological and physiological responses. Physiological arousal is diffuse, or emotionally non-specific, and determines the intensity of an emotional state, but not its quality. Cognition determines which emotion, if any, will be experienced. As described above, most theories of emotion posit that for the occurrence of an emotional state both cognition and arousal are necessary conditions, and no emotion will be experienced if either of them is missing. However, the occurrence of these two components together is not enough to experience emotion. For example, arousal and cognition of a threatening event will not create the emotion of fear, unless one labels or connects the arousal to cognition. In everyday emotional states, the cues that arouse an organism also normally provide the cognitive labels for arousal [19]. Sensory information, such visual, acoustic, tactile, olfactory, or gustatory can trigger perceptual emotions [8]. Recent neuroscience discoveries reveal that our perceptions, decisions, and behaviours are influenced by our emotions, and these influence our cognitive or logical thinking to a greater extent than once believed [11].

It is perhaps unsurprising, then, that in order to detect and measure emotion or affect we must have multiple tools at our disposal, and even then, we are limited in

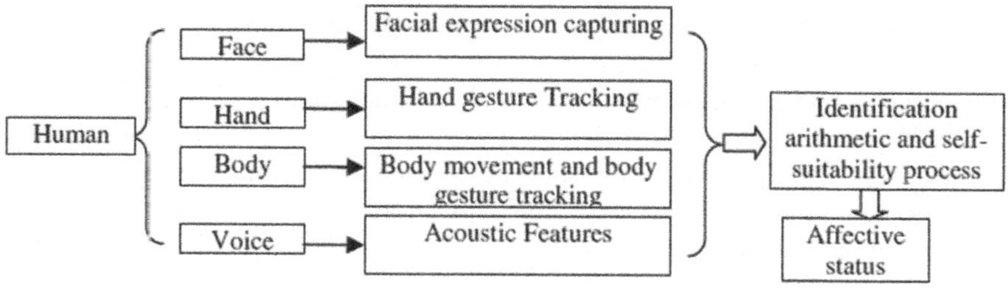

Fig. 3.1 Multi-modal input for measuring affect [21]

the data that we gather. Recent technological advances have enabled the introduction of more elaborate approaches in which the user is able to interact with the application using his/her body pose, facial expressions, actions, and physiological signals (heart beat rate, encephalogram, skin conductivity etc.).

Although people tend to polarize thoughts and emotions as if they are opposites, neurologically, the brain does not draw a hard line between thinking and feeling, and recent work into mirror neurons shows how thoughts evoke a similar affective and neuronal response to real-life experiences. Picard concludes that emotions are not a luxury but rather, are essential for rational human performance. He states that if we are to design a device that thinks in the sense of mimicking the human brain, it should both think and feel [6].

Due to the complexities of measuring emotion and affect, multimodal techniques that use multiple input measures have recently become a focus in affect/emotion recognition research. For example, data from facial expressions such as head positions, hand gestures, body postures, sounds, and physiological responses such as heart beat, pulse, sweat, and body temperature, can be used as inputs [20] (as shown in Fig. 3.1). Nevertheless, below we separate out some of the common means to measure and quantify emotional response.

3.3.1 Brain Computer Interfaces (BCIs)

A brain-computer interface device (BCI) is a hardware and software system that allows interaction with a computer directly from a user's brain through electroencephalography (EEG) which provides a means of accessing and recording neural activity (typically in a non-invasive manner with the electrodes placed along the scalp), allowing a computer to retrieve and analyse information from the brainwave patterns produced by thought [22]. Advances in technology and research allow EEG data to be processed in real time. BCIs, while detecting and creating brain activity a feedback loop, can also allow users to communicate with a computer without physical input [22]. There are many applications of BCI devices, and they have been used to restore and maintain quality of life for medically challenged people, as video game controllers for the enhancement of attention and memory, and robot control [23].

The Emotiv EPOC+ is a popular consumer-grade compact, wireless, high-resolution, portable headset BCI that is easy to set up and incorporates processing tools that can detect facial movements, emotional states, and imagined motor movement and, tracks 14 channels. Although it was developed primarily for the gaming market, and is not classified as a medical device, it has been adopted and applied to a variety of non-gaming applications due to its affordability and relative simplicity.

For example, Malete et al. [23] created a simple development platform based on the Emotiv EPOC + BCI to control a third-person 3D maze solver game where an avatar walks about the maze collecting coins and earning points, as they try to solve the maze and discover hidden treasures to score points while racing against time. The traditional navigation controls in the game are mapped to the machine learning model predictions, while a gyroscope is used for panning and rotation. Support vector machine (SVM), linear neural network (LNN) and decision trees (DT) are used to build machine learning models. The models are trained on two control states: forward and backward. Preliminary experiments in motor actions were performed on one individual and EEG recordings were taken as he walked around a room for approximately three minutes. Ten users were involved in the experiments for the motor memory approach, which consisted of two sessions involving EEG recordings for moving backwards and forward, stimulated by audio-visual modalities for a duration of 30 s each. The models were trained on two control states in order to control the game's character: (i) forward, and (ii) backward in the first stage of modelling then a more general model was developed to allow plug and play interactions with many users, without the need for re-training. Based on their results, SVM and DT classifiers were able to sufficiently model brain control with acceptable errors. They failed to generalize datasets with multiple users, however. A collection of more datasets to generalize results for multiple users using feature extraction methods is recommended [23].

Another application of BCI technology is the brain-computer musical interface (BCMI), for music creation or music interpretation. Currently there are no "plug-n-play" solutions for BCMI and it requires a special setup and specialized software. Levicán et al. [24] developed a free middleware BCMI application (sending data from an Emotiv Insight BCI in real-time), called Insight2OSC, that sends a wireless signal to an open sound control (OSC) enabled application. An example of an OSC receiving software application is MaxMSP, a widely used sound processing environment [24]. The advantages of Insight2OSC include modularity, that is, all routines are accessible from the same file, parallelism and speed, ease, and cross-platform nature.

3.3.2 Facial Expression and Eye Tracking Technologies

With human communication it has been estimated that 7% of the communication is conveyed by the explicit verbal account, 38% is conveyed by tone of voice and 55% is conveyed by facial expressions, postures, and gestures and facial expressions arise owing to a person's internal emotional states which are important means of detecting emotions [17]. Triantafyllou and Tsihrintzis [25–28] have, over the span of several years, developed the GRoup AFFect Recognition (GRAFFER) system capable of detecting emotions of groups of people. Although their work has focused on educational and entertainment applications, with an emphasis on groups of students in a diverse range of classroom settings, it can be generalized to a variety of applications. GRAFFER includes novel algorithms that, based inverted color frames, capable of: (i) iris detection, (ii) eye distance detection, and (iii) mouth opening detection. Based on these three features (iris, eye distance, and mouth opening), GRAFFER is able to make about emotional states. Ekman and Friesen devised a map of the human face, known as the facial action coding system (FACS) [29]. FACS is composed of action units (AU) which are catalogued relying on relaxation or contractions of one or more facial muscles and a facial expression is determined by multiple Action Units [30]. An example is to detect faces from videos, to process the information by extracting geometrical features, and then to produce temporal profiles of each facial movement [17]. In the FACS model there are 44 facial movements (action units) and facial expressions are generated by contractions of facial muscles, such as eyelids, eyebrows, nose, lips and skin texture, often revealed by wrinkles and bulges [31].

For instance, Microsoft's Emotion API recognizes emotions expressed by one or more people in an image, and these emotions are understood to be cross-culturally and universally communicated through particular facial expressions. There are countless electronic messages annotated with emojis, and posters such as feeling charts for children that display schematized versions of various facial expressions. The media features stories that portray facial configurations such as a scowl as being angry, or a smiling face as being happy [29].

Barrett et al. [29] raise concerns about the validity of inferring a person's emotional state from their facial movements, typically referred to as emotional expressions or facial expressions. This widespread assumption that emotional states can be read from facial expressions influences a wide range of practices, which include legal judgments, policy decisions, and national security protocols. It pervades social interactions, educational practices, and even guides the diagnosis and treatment of psychiatric illness. Technology companies are currently investing large sums to objectively read emotions in people, by detecting presumed facial expressions. Government security agents in the U.S. are trained to recognize emotions and other intentions by using facial configurations to identify and thwart terrorists. These configurations are also used to supposedly diagnose psychiatric disorders. One widely used task in autism research, the reading the mind in the eyes test, asks the subject to match photos of the upper eye region of a posed facial configuration, to specific mental state words, including emotion words [29]. Their treatment plans often include the learning and recognition of these configurations [29].

In response to these common perceptions of how facial configurations can reflect emotions, Barrett et al. [29] set out to systematically use scientific approaches. Their review focusses on six emotion categories that receive the most attention in scientific research: (i) anger, (ii) disgust, (iii) fear, (iv) happiness, (v) sadness, and (vi) surprise. They identified three key shortcomings in scientific research that contribute to misunderstandings: (i) limited reliability, (ii) lack of specificity, and (iii) limited generalization. Barrett et al. [29] propose that new research should consider sampling individuals deeply, with high dimensional measurements across many different situations and times of day. Ideally, videos of people in natural situations, could be quantified by automated algorithms for various physical features, such as facial expressions, posture, gait and tone of voice. Eye tracking could assess gaze and attention.

Electromyogram sensors measure both the visible and subtle facial muscle movements by recording the muscles' electrical activity. The face is a rich expressive area of communication and when communicating with others, even though we may not necessarily be aware of it, our attention is placed on their face and we are well perceptive of subtle facial expressive signs [2]. Branco and Encarnação, [2] conducted a study to develop an understanding of the relevance of the user's facial expressions in identifying problematic events during the course of a typical computer interaction scenario. The participants used Microsoft Word 2000 to format a page of typed text to resemble a target printed page which was presented to them while they were video monitored. Participants were required to format text, change colours, draw lines, create columns, insert pictures, create and format tables, insert word-art and textboxes, edit the footer of the page. They were also asked to work until the task was completed or, if they had difficulties with any of the components of the task, to continue and keep working until they felt they had achieved as much as they could. Their initial results suggest that facial expressions during problematic events are exhibited during the interaction.

An eye tracker (hardware and software) monitors a participant's visual attention via eye-movement data. Eye movements are essential to cognitive processes because they focus the participant's visual attention to the parts of a visual stimulus that are processed by the brain and thus, camera-based eye tracking can be used for emotion detection [32]. Eye activity, and eye movement, pupil dilation, and pupil visibility, could be used to complement the automatic recognition of human emotional states (and it can be complemented with other modalities such as EEG to improve its accuracy) [32]. Research has shown that pupil dilation is sensitive to light in addition to memory load, cognitive difficulty, pain and emotional state [32]. Extreme pupil dilation occurs for interesting or pleasing stimuli images [33], while it has been shown that pupil size is significantly larger during both emotionally negative and positive stimuli than during neutral stimuli [34]. Eye tracking provides an indication of learner interest and focus, and can be used alongside animated agents (software agents that have intelligence and knowledge about the learner, in order to personalize and focus the instructional strategy, and character agents who represent people).

3.3.3 Galvanic Skin Response

Galvanic skin response (GSR), also known as skin conductance (SCR) or electrodermal response (EDR) measures the change in skin resistance and is the basis to the original "lie detector" testing employed by policing (today, lie detection is often much more sophisticated and uses multiple methods of input including heart rate, respiration and even fMRI). Skin conductance is the response of any bio-medical system to the electrical properties of skin due to variation in physiological and psychological conditions and this change is caused by the degree to which a person's sweat glands are activated (stress tends to make the glands more active and this lowers the skin's conductance) [8]. The electrodes, which correspond to the GSR sensor, are tightly fastened to the index finger and the middle finger. The device is accurate, hand-held, convenient, and includes sensitivity controls, which vary its responses to changes in skin conductivity. In a study conducted by Vijaya and Shivakumar, [8], 75 participants fitted with a GSR sensor were shown movie clips (which included both audio and visual stimuli) to evoke four prominent emotions: happy, surprise, disgust and fear, with the goal of changing their emotional state by external influences. Between the clips, participants were provided sufficient time to return to a relaxed baseline, or a neutral emotion. Results indicate that recognition of most of the emotional states using GSR is above 80%, and it is easier to evoke and identify positive emotions such as happiness, than the negative emotions of fear and disgust.

3.3.4 Multimodal Input Devices

As discussed above, a single input often provides unreliable results in attempting to accurately detect and measure affect. As such, it is often desirable to employ multiple tools and techniques simultaneously. Some recent devices have provided for easier capture and synchronization, including for instance the Microsoft Kinect. The Kinect vision-based sensor captures, at the same time, both 2D and 3D information with one RGB camera and two 3D depth sensors under ambient light conditions, capturing movement and body position along with facial expression. Wei and Jia, [17] investigated 3D facial emotion recognition using the Kinect sensor in real time. Their overall goal was to apply it to robotics for more robust 3D facial expression recognition. They proposed three differences in their study as compared to existing research at the time. First, they used depth information, which is a more advanced processing strategy as opposed to feature extraction, which relies on texture and appearance information. Second, for improving the facial feature presentation, dynamic and static features were used. Third, they used two databases, the RGB-D database and the KinectFaceDB database, with eight emotions. The first database, the RGB-D database, includes five emotional expressions: (i) neutral, (ii) smile, (iii) sad, (iv) yawn, and (v) angry. The second database, the KinectFaceDB database, contains the other three expressions: (i) smile, and (ii) open mouth, and (iii) neutral. They applied a machine learning (ML) algorithm on their training set data. They found little variability in the eyes in the different facial expressions and conclude that in order to recognise facial expressions, movement of the mouth or mouth behaviours are required [17].

Surveillance systems can be used in the early detection of aggressive and provocative actions in institutions and public places, they have the potential to mitigate such behaviour. Patwardhan and Knapp, [20] developed an affect-aware surveillance system that employs a fusion strategy that combines the detection of facial expressions, body postures, head movements, hand movements and speech with an infrared Microsoft Kinect sensor. They conducted an experiment that included five participants who were asked to enact anger or aggression through gestures and language, which included raising their voice and tone in front of the Microsoft Kinect sensor. Participants displayed non-aggressive emotional gestures for 21 sessions (three acts for each of the seven emotional states). Seven classes were tracked which included six basic emotions and one neutral state. The system includes three classifiers, one relying on support vector machines (SVMs), one on the rule-based machine learning method and the last relying on SVMs combined with the rule-based machine learning method. These learning algorithms analysed feature vectors (tracking joint locations), for example, the position of the hand was defined/tracked by co-ordinates of the shoulder, and 60 tracked points on the facial region included the lower and upper lips, lower and upper eye lid, left and right cheek, chin, and tip of nose [20]. Two experiments were also conducted to examine the performance of the system, and more specifically, the accuracy of each of the underlying classifiers. Results of the first experiment (that considered the SVM classifier only), showed that the emotions of fear and surprise resulted in a classification rates of 83.3% and 81.7% respectively, while the emotions of anger, disgust, and happy resulted in recognition rates of 74.4%, 73.2%, and 76.8% respectively.

Using rule-based classification, the overall performance of all the features was 93.51%. The emotions with high precision were anger (99.6%), sad (99.4%), disgust (98.3%) and happy (91.1%). The precision for surprise (86.1%) and fear (85.5%) was low compared to the other 4 emotions. With both, SVM and rule-based classification, the "most recognized emotions were disgust (99.6%), surprise (99.2%), fear (98.8%) and anger (91.7%). Sad was misclassified with fear (16.2%). Happy was misclassified with surprise (13.9%)" [20]. The classification, precision and recall rates showed improvement by combining the classifier and rule-based recognition. Based on these results, Patwardhan and Knapp [20] conclude that a Microsoft Kinect-based infrared surveillance system is feasible for some emotion detection and can be used in early detection of aggressive behaviour.

3.3.5 Emotional Speech Recognition and Natural Language Processing

Emotional speech recognition (ESR), multi-keyword spotting (MKS) and video feature extraction (VFE) are used to analyse speech and video input of users. This allows for multimodal input and the (affective input) data is streamed (using communication protocols) and further processed to create an overall mood of the session [35]. Combined with processing what is said (using natural language processing, or NLP), this tool can prove somewhat reliable [4]. The original Turing test was designed via text, and therefore sensory expression was not taken into consideration, but it was clear that emotions could be perceived in text alone [6]. The ability to accurately identify, reflect on and detect subjective nuances in written text is important. It influences real-world text analysis applications such as targeting marketing campaigns to users with specific likes and dislikes or identifying online antisocial behavior such as cyberbullying [36].

With NLP, there is a need to differentiate between subjective affective terms, such as affect, feelings, emotions, sentiments, and opinions. Since subtleties and nuances expressed make understanding difficult, there is a subpar detection of the terms in text. NLP applications may benefit from clear objective definitions. Examples include marketing campaigns, which target users with specific likes and dislikes using keywords designed to elicit emotion, or in identifying online antisocial behaviour.

Recognizing the lack of tools or databases to be used as emotional linguistic resources, Valitutti et al. [4] developed Wordnet-Affect as a tool for NLP applications to understand affect. Wordnet-Affect was developed by building upon the lexical knowledge base WORDNET, through a selection and labelling of the affective concepts (represented by sets of synonyms). A lexical structure was built by studying the terms related to emotions, and is based on lexical semantics that state it is possible to infer emotion properties from the emotion words (newspaper texts and dictionaries were used to determine or collect the emotional words) [4]. The semantic contexts are then fixed (e.g., pure emotion terms, personality trait terms, physical and cognitive state terms, etc.), and factorial analysis or multidimensional scaling was used to extract a set of affective dimensions [4]. WORDNET-AFFECT allows such systems to perform reasoning not only on the "word" level but on the "sense" level (through the synsets linked to each word in the resource, allowing for the potential in the performance of the affective understanding from texts.

Wordnet-Affect was later extended with the OpenMind database, which contains a set of common sentences, to create OpenMind-Affect. OpenMind-Affect contains sentences, patterns, and parsing trees containing affective concepts, and this information allowing for the collection of contextual information, such as events that cause specific emotions [26]. Using contextual information can increase affective knowledge of lexical items and allow for the possible identification of sentences containing words with an affective meaning and to select a subset of OpenMind.

More recently, Google's word2vec neural text processing tool—combined with steady increases in computing power, have broadened the potential for natural language processing far beyond the research lab or supercomputer [37]. The vector model approach, where each word is given a series of scores that position it in an arbitrary space, has gained enormous traction in the past several years.

3.4 Application Examples

It is clear that the tools for the measurement and quantification of emotional expression are improving, especially as they incorporate elements derived from machine learning. As such, the cost of undertaking research (whether private, public, or corporate) into capturing and analysing emotion has become greatly reduced, along with the difficulties previously inherent in many of the earlier generation of tools. We conclude our chapter by highlighting some of the current uses of affective computing across a variety of fields.

3.4.1 Entertainment

An entire field of affective games, or affective elements in games has gradually been gaming steam after years of experimental attempts to incorporate affective detection in game input devices. Affective gaming today can not only to extract the player's commands by their speech and gestures but also by the player's behavioural cues and emotional states and can adjust their game accordingly, in order to ensure more realistic and satisfactory gameplay experiences [38]. Similarly, with the growth of virtual and augmented reality systems, the integration of sensors has enabled more natural interactions, although many of these still remain in the experimental, research and art-based gaming areas. Where affective computing has perhaps been most useful is in educational and simulation-based training.

Argasiński et al. [39] developed a virtual reality-based serious game simulation for fire-fighter training with affective feedback. It involved collecting physiological data from participants, in stress-inducing situations where there is a time pressure. The game simulates various types of traffic accidents with a variety of victims, and the users (firefighters) are equipped with authentic fire fighter equipment. Various physiological responses are measured (e.g., heart rate, skin conductance, muscle activity). Through the use of game mechanics influenced by narrative and aesthetic design, the developers of the VR simulation have a direct impact on the player's experience. The rules stated as verbs, such as "open" the door, are translated into game mechanics and are the most special element in designing game software.

Quesnel et al. [40] have been exploring the phenomenon of "awe" or "chills" in virtual reality. They developed 'Earthgazing', an immersive environment simulation that provides a visualization of different places on earth. In a study using Earthgazing and a control (a different informational VR game of nature scenes designed to not produce the same affect), they examined feelings of peak positive affect while the subject was interacting in an immersive VR environment. Both applications employed an HTC Vive head mounted display (HMD) and an HCT Vive hand controller. Along the time series (TS), of events occurring with the Virtual Reality display, the facilitator annotated the comments by the participants of their affective states. The time series displayed the participant bio signals, and the emotions were also coded after the sessions concluded. Machine learning was used on the data (classification using a neural network) allowing the system to adjust to the user's experience of peak experience. By creating a trained model, the VR system can become bio-reactive to adjust the user's environment and change their personal experience. For example, if users are predicted to experience fear, the system integrates calming audio-visual stimuli and if the user is possibly experiencing awe, the system may respond with a crescendo of music and aesthetic beauty [40].

The research involved the collection of skin conductance and goose bump physiological data, as well as labelled-recalled events from participants for training data, but is still in the developmental stage. Once these data are collected, they will be released as open source for public use. Quesnel et al. [40] hope the work will be utilized by researchers, designers, and artists for evaluation and creation of bio-responsive, personalized VR environments. As a type of emotive media, Earthgazing, which uses bio-feedback sensors, can create a pathway for interactive applications to discover emotional states. Such data could be particularly useful to help differentiate between profound affective states that may have similar bio-signal features in at least one sensor modality [40].

3.4.2 Chatbots

A chatbot is a software agent used to interact between a computer and a human using natural language. Just as people use language for human communication, a chatbot uses natural language to communicate with human users. A chatbot can process language input, such as a question from the user, by using NLP [41]. Current chatbot frameworks are able to maintain and support explicit highlights and settings of the discussions empowering them to have human interaction in real-time surroundings [42]. Wan et al. [43] developed a sentiment-analysis network (SAN), using chatbot technology that responds to input from the user. It differs from the typical existing chatbots that only interact using text and voice. In contrast, their chatbot is intended to have human interaction in real-time surroundings. This work uses a long short-term memory network- (LTSM-) based emotional 3D module chatbot, "Midoriko". A LSTM affords an excellent interactive interface and ductility. Features of the system include the ability to choose characters, scenes, and sound effects. The system can also be extended using immersive technologies (virtual and augmented reality) [43]. The system calculates an emotional score from the user's response and sends it to the 3D module, which controls an avatar.

Wan et al. [43] believe that chatbots created with artificial intelligence may alleviate the problems caused by alienation in society. In addition to text and sound, a chatbot can exhibit expressions and body movements through graphical interfaces such as role-playing. It is not just a task-orientated AI assistant, but can serve as a friend or a companion. Such therapeutic-based interventions are already being deployed. Avatars can engage users in conversations that that can lead to effective therapies for improving social skills, introspection (self-understanding) and empathy.

For instance, people who are diagnosed with different forms of social communication disorder (SCD) typically have problems with social understanding, social interaction, and pragmatics, that is, the role of context on a word's meaning [44]. Johnson et al. [45] report on the design and evaluation of an affective avatar intended to help those with SCD. The avatar is designed to address the emotional state and cognitive process to assist people in the management of emotions, and supports cognitive function and theory of mind, which involves the capacity to deduce the mental state of another person. The avatar's design is based on neurological function, and mirror neurons, which are involved in the acquirement of knowledge and building empathy. Since the intent of the avatar is to help the user engage in real life social communication, it was decided to provide a humanoid appearance capable of expressing various emotions including happiness, sadness, anger, surprise, fear, an idle state consisting of blinking and a neutral state. The avatar transitions between these states, depending on actions and interactions of the user. Based on the results of a study conducted to examine how well the avatar is able to provide users the ability to experience and exhibit empathy, they conclude that avatar can evaluate and assist people with SCD.

3.4.3 Medical Applications

Aside from therapeutic chatbots, other medical applications are being developed to assist with interventions in a variety of maladies. For instance, Siegmund et al. [46] developed Talis, a wearable user-interface shaped like a talisman and worn around the neck, to support people suffering from mild to moderate depression. Positive moments are detected automatically through speech (diary-like speech messages to oneself or sound landscapes of a favourite place), and recorded and the collected data can be used to create a mood image, which consists of the detected emotions, evaluated on a scale from purely negative to purely positive over a period of time. This image provides the therapist with a visual representation of emotional progression. Using emotion recognition technology, haptic communication is initiated when a negative mood is detected over a long period of time [46]. An Emo-voice framework, which classifies emotions based on frequency, volume and the relationship between tone and noise (HNR) in speech, was used to train the Talis AI system. Training involved the use of existing labelled databases as well as the researchers' own voice data. A support vector machine (SVM) algorithm was used on the data collected by the device and training consisted of classifying the emotional states as joy, satisfaction, anger and frustration. The emotion-recognition algorithm analyzes samples with a length of a few seconds. After a certain number of negative samples, the wearable will play a positive recording. Phases of depression that may be linked to the patient's personal situation can be discussed and analysed during a therapy session, and together the patient and therapist can work on ways to avoid negative phases. Access to the data occurs in conjunction with a therapist to avoid negative reflections and a worsening of the patient's condition.

Alepis and Virvou [47] developed an authoring tool that can be used by a medical educators to construct animated medical tutor agents (a cartoon doctor) with emotional interaction capabilities, capable of who can responding emotionally to medical students during their interactions and thus make e-learning more human-like. The tutor's behavior relies on the Ortony, Clore, and Collins (OCC) model of emotion [14], and can move around the information presented to the trainee, show the trainee a theory that they are expected to read and can show patience while the student reads a theory, and boredom if the student is not responding to the system amongst others. The tutor can also capable of speaking to the trainee and their speech can be parametrized to convey emotion (e.g., anger and enthusiasm).

There are, of course, many other applications for affective computing than the three we have covered here, and the field continues to grow at a rapid pace as costs and difficulties with both the machine learning side of processing data and the collection of affective data itself becomes more affordable (for a more complete overview, see for instance Novak et al. [48].

3.5 Conclusions

In this chapter, we have introduced the basics and current state of the art of affective computing. There are still many hurdles faced when it comes to affective computing. As described above, the first hurdle is the simple fact that scientists do not yet agree on what emotion is, what where or how it is formed, and why. Categorizing emotion and affective response is also a source of much contention and the current models are proving in many cases to be ineffective. With respect to the available tools, there are also disagreements on the best way to use them, for instance, on how emotions can be interpreted from facial data [29], the datasets used to train neural networks, noise in the data, and many other issues beset the current measurement of affect. Multiple sources of input are clearly superior to a single input mode when it comes to detection accuracy (e.g., see Virvou et al. [49]), yet this increases the costs and data processing requirements involved. Although tools are getting easier in terms of their use, the data analysis required still involves expertise and considerable training.

There are also, of course, ethical issues that arise from the development and use of affective technologies, particularly when it comes to the use of such systems in surveillance and invasions of privacy. The many problematic approaches to artificial intelligence and its datasets—often well documented due to the biases of tech companies in terms of gender and race—raise similar issues when it comes to the data used for affective computing. Some research has been done to develop therapies for the disabled on people who do not have that disability [45]. Therapy or medical tools that may misdiagnose or alert authorities when they sense danger may be mistaken and the subsequent actions may interfere with a person's life in a negative way.

There are also difficulties with most of the methods used to measure emotion. With respect to natural language processing, it is difficult to argue for a correct model to use in labelling emotions in text since they are all founded on different psychological the theories of emotion. Munezero et al. [36] argue that NLP approaches need to be extended, such as in the cultural domain since the choice of words and their intended meaning are personally, contextually, culturally, and socially dependent. When it comes to facial recognition, most machine learning algorithms use the Ekman model which is the facial action coding system (FACS). Can we be confident that Ekman's basic prototypic emotional displays (FACS), which include the basic emotions of happiness, sadness, fear, disgust, surprise, and anger are accurate? After extracting suitable features from the sensor data, various machine-learning and data mining techniques are applied to automatically recognise participants' emotions, such as anger and happiness; mood, such as positive and negative mood; or personality traits, such as neuroticism or extraversion [50]. Given that emotional expression can be culturally contextualized, how reliable is the extracted data? Is there a bias based on the researchers' cultural expectations?

What are the consequences of developing affective systems? Can we create computers that can have some autonomous agency over us, as well as exhibit humour and creativity and never use emotional actions to harm us [6]? "Who wants a computer to be able to "feel angry" at them? To feel contempt for any living thing" [6]? A larger issue is of course privacy rights: who owns our emotional response data if we are watching a film, listening to a piece of music, and our responses are recorded. Should there be public disclosure of affect databases which contain multimodal affect data [50]?

Despite the many difficulties and issues raised when it comes to the current state of the art in affective computing, as also shown in this chapter, affective computing offers many intriguing possibilities that can improve our lives: companions that can support us, therapists that can heal us, games that can entertain or teach us, chatbots that can help us navigate a difficult system, and so many more applications that mean we should not abandon affective computing simply because there are potentials for abuses.

Acknowledgements The financial support of the N*atural Sciences and Engineering Research Council of Canada* (NSERC) and the *Social Sciences and Humanities Research Council of Canada* (SSHRC), is gratefully acknowledged.

References

1. A.M. Turing, Computing machinery and intelligence. Mind **LIX**(236), 433–460 (1950)
2. P. Branco, L.M. Encarnação, Affective computing for behavior-based UI adaptation, in *International Conference on Intelligent User Interfaces* (Funchal, Madeira, Portugal, 2004)
3. B. Kratzwald, S. Ilić, M. Kraus, S. Feuerriegel, H. Prendinger, Deep learning for affective computing: text-based emotion recognition in decision support. Decis. Support Syst. **115**, 24–35 (2018)
4. A. Valitutti, C. Strapparava, O. Stock, Developing affective lexical resources. PsychNology J. **2**(1), 61–83 (2004)

5. K. Amara, N. Ramzan, N. Achour, M. Belhocine, C. Larbas, N. Zenati, Emotion recognition via facial expressions, in *2018 IEEE/ACS 15th International Conference on Computer Systems and Applications (AICCSA)* (2018), pp. 1–6
6. R.W. Picard, Affective computing-MIT media laboratory perceptual computing section. Technical Report No. 321 (Cambridge, MA, 2139, 1995)
7. R.C. Balabantaray, M. Mohammad, N. Sharma, Multi-class twitter emotion classification: a new approach. Int. J. Appl. Inf. Syst. **4**(1), 48–53 (2012)
8. P.A. Vijaya, G. Shivakumar, Galvanic skin response: a physiological sensor system for affective computing. Int. J. Mach. Learn. Comput. **3**(1), 31 (2013)
9. W.B. Canon, The James-Lange theory of emotions: a critical examination and an alternative theory. Am. J. Psychol. **39**(1/4), 106–124 (1927)
10. M. Saraiva, H. Ayanoğlu, Emotions and emotions in design, in *Emotional Design in Human-Robot Interaction.* (Springer, Cham, 2019), pp. 57–70
11. S.L. McShane, S.L. Steen, *Canadian Organizational Behaviour*, 8th edn. (2012)
12. K.M. Heilman, Emotion and the brain: a distributed modular network mediating emotional experience, in *Neuropsychology*. ed. by D. Zeidel (San Diego, CA, Academic Press, 1994), pp. 139–158
13. D.G. Myers, *Theories of Emotion. Psychology*, 7th edn. (Worth Publishers, New York, NY, 2004)
14. G.L. Clore, A. Ortony, Psychological construction in the OCC model of emotion. Emot. Rev. **5**(4), 335–343 (2013)
15. K.R. Scherer, What are emotions? And how can they be measured? Soc. Sci. Inf. **44**, 695–729 (2005)
16. B. Plotkin, *Wild Mind: A Field Guide to the Human Psyche* (New World Library, Novato, CA, 2013)
17. W. Wei, Q. Jia, 3D facial expression recognition based on Kinect. Int. J. Innov. Comput. Inf. Control **13**, 1843–1854 (2017)
18. J.R. Fontaine, K.R. Scherer, E.B. Roesch, P.C. Ellsworth, The world of emotions is not two-dimensional. Psychol. Sci. **18**(12), 1050–1057 (2007)
19. R. Reisenzein, The Schachter theory of emotion: two decades later. Psychol. Bull. **94**(2), 239–264 (1983)
20. A. Patwardhan, G. Knapp, Aggressive actions and anger detection from multiple modalities using Kinect. arXiv preprint arXiv:1607.01076 (2016)
21. J. Tao, T. Tan, Affective computing: a review, in *International Conference on Affective Computing and Intelligent Interaction* (Beijing, China, 2005), pp. 981–995
22. M. Lang, Investigating the Emotiv EPOC for cognitive control in limited training time, Honours Report (Department of Computer Science, University of Canterbury, 2012)
23. T.N. Malete, K. Moruti, T.S. Thapelo, R.S. Jamisola, EEG-based control of a 3D game using 14-channel Emotiv Epoc+, in *2019 IEEE International Conference on Cybernetics and Intelligent Systems and IEEE Conference on Robotics, Automation and Mechatronics* (Bangkok, Thailand, 2019), pp. 463–468
24. C. Levicán, A. Aparicio, V. Belaunde, R.F. Cádiz, Insight2osc: using the brain and the body as a musical instrument with the Emotiv Insight, in *International Conference on New Interfaces for Musical Expression*, (2017), pp. 287–290
25. A.M. Triantafyllou, G.A. Tsihrintzis, Group affect recognition: completed databases & smart uses, in *ACM 3th International Conference on E-Education, E-Business and E-Technology (ICEBT)* (2019) pp. 38–42
26. A.M. Triantafyllou, G.A. Tsihrintzis, Group affect recognition: optimization of automatic classification, in*Springer 12th Joint Conference on Knowledge-Based Software Engineering (JCKBSE)* (2018), pp. 189–196
27. A.M. Triantafyllou, G.A. Tsihrintzis, Group affect recognition: Visual - facial collection, in *IEEE 29th International Conference on Tools with Artificial Intelligence (ICTAI)* (2017), pp. 677–681

28. A.M. Triantafyllou, G.A. Tsihrintzis, Group affect recognition: evaluation of basic automated sorting, in *IEEE 9th International Conference on Information, Intelligence, Systems and Applications (IISA)* (2018), pp. 1–8
29. L.F. Barrett, R. Adolphs, S. Marsella, A.M. Martinez, S.D. Pollak, Emotional expressions reconsidered: challenges to inferring emotion from human facial movements. Psychol. Sci. Public Interest **20**(1), 1–68 (2019)
30. S. Tornincasa, E. Vezzetti, S. Moos, M.G. Violante, F. Marcolin, N. Dagnes, L. Ulrich, G.F. Tregnaghi, 3D facial action units and expression recognition using a crisp logic. Comput. Aided Des. Appl. **16**, 256–268 (2019)
31. B. Fasel, J. Luettin, Automatic facial expression analysis: a survey. Pattern Recognit. **36**(1), 259–275 (2003)
32. S. Alghowinem, M. AlShehri, R. Goecke, M. Wagner, Exploring eye activity as an indication of emotional states using an eye-tracking sensor, in *Intelligent Systems for Science and Information*, eds. By L. Chen, S. Kapoor, R. Bhatia (Springer, 2014), pp. 261–276
33. P.A. Chiesa, M.T. Liuzza, A. Acciarino, S.M. Aglioti, Subliminal perception of others' physical pain and pleasure. Exp. Brain Res. **233**(8), 2373–2382 (2015)
34. H.I. Liao, M. Yoneya, S. Kidani, M. Kashino, S. Furukawa, Human pupillary dilation response to deviant auditory stimuli: effects of stimulus properties and voluntary attention. Front Neurosci. **10**, 43 (2016)
35. S.W. Gilroy, M. Cavazza, R. Chaignon, S.M. Mäkelä, M. Niranen, E. André, T. Vogt, M. Billinghurst, H. Seichter, M. Benayoun, E-tree: emotionally driven augmented reality art, in *16th ACM International Conference on Multimedia* (Vancouver BC, Canada, 2008), pp. 945–948
36. M. Munezero, C.S. Montero, E. Sutinen, J. Pajunen, Are they different? Affect, feeling, emotion, sentiment, and opinion detection in text, IEEE Trans. Affect Comput. **5**(2), 101–111 (2014)
37. G. Goth, Deep or shallow, NLP is breaking out. Commun. ACM **59**(3), 13–16 (2016)
38. I. Kotsia, S. Zafeiriou, S. Fotopoulos, Affective gaming: a comprehensive survey, in *IEEE Conference on Computer Vision and Pattern Recognition Workshops* (Portland, OR, USA, 2013), pp. 663–670
39. J.K. Argasiński, P. Węgrzyn, P. Strojny, Affective VR serious game for firefighter training, in *Workshop on Affective Computing and Context Awareness in Ambient Intelligence (AfCAI)* (Valencia, Spain, 2018), p. 43
40. D. Quesnel, S. DiPaola, B.E. Riecke, Deep learning for classification of peak emotions within virtual reality systems, in *International SERIES on Information Systems and Management in Creative eMedia (CreMedia)*, (2017/2), (2018), pp. 6–11
41. V. Sharma, M. Goyal, D. Malik, An intelligent behaviour shown by chatbot system. Int. J. Res. Eng. Technol. **3**(4), 52–54 (2017)
42. S. Patil, V.M. Mudaliar, P. Kamat, S. Gite, LSTM based ensemble network to enhance the learning of long-term dependencies in chatbot. Int. J. Simul. Multidiscip. Des. Optim. **11**:Article 25 (2020)
43. Y.T. Wan, C.C. Chiu, K.W. Liang, P.C. Chang, Midoriko Chatbot: LSTM-based emotional 3D avatar, in *2019 IEEE 8th Global Conference on Consumer Electronics* (Osaka, Japan, 2019), pp. 937–940
44. A. Papafragou, Pragmatic development. Lang. Learn. Dev. **14**(3), 167–169 (2018)
45. E. Johnson, R. Hervás, G. López, C. de la Franca, T. Mondéjar, S.F. Ochoa, J. Favela, Assessing empathy and managing emotions through interactions with an affective avatar. Health Inf. J. **24**(2), 182–193 (2018)
46. D. Siegmund, L. Chiesa, O. Hörr, F. Gabler, A. Braun, A. Kuijper, Talis—A design study for a wearable device to assist people with depression, in *2017 IEEE 41st Annual Computer Software and Applications Conference July 4–8*. (Italy, Turin, 2017), pp. 543–548
47. A. Alepis, M. Virvou, Automatic generation of emotions in tutoring agents for affective e-learning in medical education. Expert Syst. Appl. **38**(2011), 9840–9847 (2011)

48. D. Novak, G. Chanel, P. Guillotel, A. Koenig, Guest editorial: toward commercial applications of affective computing. IEEE Trans. Affect Comput. **8**(2), 145–147 (2017)
49. M. Virvou, G.A. Tsihrintzis, E. Alepis, I.-O. Stathopoulou, K. Kabassi, On the use of multi-attribute decision making for combining audio-lingual and visual-facial modalities in emotion recognition, in *Tsihrintzis GA*. ed. by M. Virvou, L.C. Jain, R.J. Howlett, T. Watanabe (Intelligent Interactive Multimedia Systems and Services in Practice, Springer, 2015), pp. 7–34
50. E. Politou, E. Alepis, C. Patsakis, A survey on mobile affective computing. Comput. Sci. Rev. **25**, 79–100 (2017)

Chapter 4
A Database Reconstruction Approach for the Inverse Frequent Itemset Mining Problem

Panteleimon Krasadakis, Evangelos Sakkopoulos, and Vassilios S. Verykios

Abstract Preservation of sensitive knowledge which is extracted from transactional data of individuals is an increasingly popular field in the data mining research community. In itemset-based mining in particular, a key approach is based on the well-known Apriori algorithm. This paper proposes a novel hiding scheme that produces a sanitized database based on the ideal borders that are computed through the application of a constraint-based frequent itemset mining algorithm. The goal of the proposed approach is to improve at large on computational aspects of the hiding methodology, and in particular to examine how the inverse frequent itemset mining problem can be used in data hiding that ensures privacy protection.

Keywords Privacy techniques · Frequent itemset hiding · Constraints data mining · Database reconstruction · Inverse frequent itemset mining

4.1 Introduction

The rapid growth of information stored in different datasets necessitates a subsequent progress in data mining techniques, whose usefulness become more prominent than ever before with the constantly increasing size and complexity of these datasets. Many researchers, companies and governments on a regular basis process large amounts of data in order to extract knowledge or patterns, or any other kind of valuable information. A common tactic of added value in this context which manifests itself when information is shared among companies or organisations, brings a lot of advantages, but it may also result in the danger of unauthorized access to sensitive information that the users do not consent to share.

P. Krasadakis · E. Sakkopoulos (✉)
University of Piraeus, Piraeus, Greece

V. S. Verykios
Hellenic Open University, Patras, Greece

M. Virvou et al. (eds.), *Advances in Artificial Intelligence-based Technologies*, Learning and Analytics in Intelligent Systems 22,
https://doi.org/10.1007/978-3-030-80571-5_4

As a result, the development of privacy preserving techniques for the processing of shared information has become essential. Clifton and Marks were the first to introduce the concept of privacy centered information disclosure in data mining [1] and to propose some initial solutions for sensitive knowledge protection. These led to Privacy Preserving Data Mining [2] becoming a more prominent field of research among the research community. The two main problems addressed in PPDM are the protection of sensitive data (items) and the other is the protection of sensitive rules contained in that data. The end goal in both of these scenarios is to provide "sanitized" data, data cleared from protected information. In the most recent years, it has become even more so, due to numerous well-known security and privacy breaches and the establishment of strict regulations such as the GDPR [3] (General Data Protection Regulation) that pose data minimization rules and consent based sharing of information only.

Multiple approaches have been researched and developed to protect the sensitive knowledge by sanitizing data and they are commonly known as knowledge hiding techniques. We focus on the major subdomain of knowledge hiding, the **Frequent Itemset Mining** [4]. This approach aims to sanitize a database from a set of sensitive frequent itemsets, in such a way that the new database prevents the sensitive knowledge from being data mined, while trying to maintain the best data quality in that new database (in comparison to the original database). Cafaro and Pulimeno in [5] thoroughly review the most important techniques and algorithms used in Frequent Itemset Hiding.

As the above survey suggests numerous different algorithms and techniques have been introduced alongside variations of the problem. We want to highlight two of them. The first one is the parallel implementation of such algorithms in order to boost their efficiency and scalability. We proposed such an algorithm in [6] and the initial tests and evaluation are promising. The second technique is Mining problem and dataset reconstruction. Its initial formulation and complexity has been studied by Mielikainen [7]. For the Inversion problem, the goal is to compute a database from a given set of Itemsets in such a way that they remain frequent after privacy protection is put in place.

The purpose of the work presented in this paper is to combine the extensive research in these two fields and propose a solution to the Inverse Frequent Itemset Hiding Problem. We introduce the Border based revision in the dataset reconstruction algorithm. In this way, we efficiently produce a dataset that has been sanitized from sensitive information.

In Sect. 4.2 we present the most significant related work in the area of Frequent Itemset Hiding Problem that inspired and helped to the creation of this paper. Section 4.3 states the Frequent Itemset Hiding Problem, the constraint-based algorithm that for border revision, the Inversion problem and the dataset reconstruction algorithm, all essential to our approach. Section 4.4 presents our combined analytic approach to the constraint-based hiding model. Section 4.5 concludes this paper and suggests future steps in our research and of this field in general.

4.2 Related Work

In the field of Frequent Itemset Mining (FIM), no other algorithm is used more than the Apriori algorithm. Its main function is to generate frequent itemsets and their association rules from a transactional database. The frequent itemset hiding problem has been proven NP-hard by Atallah et al. [8] and most hiding methodologies are based on their inaugural work and their greedy algorithm that tries to change transactions in order to hide the sensitive frequent itemsets, but it does not offer any guarantees on the data quality of those changed transactions.

Dasseni et al. [9] advanced the research by generalizing the problem of hiding sensitive knowledge by proposing algorithms for hiding association rules other than the usual frequent itemsets. Gkoulalas -Divanis and Verykios [10] proposed and developed linear programming techniques to optimally solve the problem, while also applying the border-based principle. Another novel approach proposed by Stavropoulos et al. in [11] translates the FIM problem to minimal transversals of a hypergraph.

The Inverse Frequent Itemset Hiding Problem (IFIM) [7] has been researched as well, though not as much as the original problem. Wang et al. in [12] prove that the complexity of the inverse frequent itemset mining is NP-complete and propose an approximate algorithm. Guo proposed in [13], a FP-tree based algorithm for a reconstruction based framework that aims to hide association rules. Guzzo et al. present a solution to the IFIM problem in [14] that combines infrequency constraints and linear programming. The most recent works include the privacy preserving database reconstruction apporach of Li et al. in [15] and the extension of the problem for realistic datasets by Sacca et al. in [16].

The most recent direction of itemset hiding problem is focused in parallel implementations and variations of Apriori algorithm. The MapReduce parallel framework works as the foundation of those approaches that try to improve their execution efficiency for the large amounts of data that exist nowadays. The state-of-the-art propositions currently are the improved Apriori on Spark by Gao et al. in [17] and our recent work in [6] that implements the constraint and border based approach with the MapReduce framework.

In the next sections, we will introduce an effective database reconstruction approach for the hiding problem of sensitive data focusing on privacy protection. First, we briefly introduce the frequent itemset hiding problem and its inverse variant.

4.3 Problem Definition

In this section we will shortly describe the basics of the Frequent Itemset Hiding Problem and its Inverse counterpart.

4.3.1 Frequent Itemset Hiding Problem

Let $\mathcal{I} = \{i_1, i_2, \ldots, i_n\}$ be a set of distinct literals (items) and X be a k-itemset (i.e., $|X| = k$). A transaction database $\mathcal{D}$ is a set of transactions between the itemsets. The support count of an itemset X in $\mathcal{D}$, denoted by $\text{supc}_{\mathcal{D}}(X)$, is the cardinality of the set of transactions supporting X. Equivalently, we define the *support* of an itemset X, denoted by $\text{sup}_{\mathcal{D}}(X) = \frac{\text{supc}_{\mathcal{D}}(X)}{|D|}$, as the fraction of the support count over the total count of transactions in the database $\mathcal{D}$.

Given a user-specified support threshold σ, we call an itemset X *frequent* in $\mathcal{D}$ iff $\text{sup}_{\mathcal{D}}(X) \geq \sigma$. Let $\mathcal{F}_{\mathcal{D}}^{\sigma}$ be the collection of σ-frequent itemsets in $\mathcal{D}$ and $\mathcal{S}_{\mathcal{D}}^{\sigma} \subset \mathcal{F}_{\mathcal{D}}^{\sigma}$ be the set of sensitive and frequent itemsets that need to be hidden. Based on the above, the aim is to transform $\mathcal{D}$ to $\mathcal{D}'$ in such a way that we minimize (a) the number of the sensitive itemsets that are not hidden and (b) the number of hidden non-sensitive itemsets.

Let the positive border of $\mathcal{F}_{\mathcal{D}}^{\sigma}$ be $\mathcal{B}d^{+}(\mathcal{F}_{\mathcal{D}}^{\sigma}) = \{X \in \mathcal{F}_{\mathcal{D}}^{\sigma} \mid X \subset Y \text{ implies } Y \notin \mathcal{F}_{\mathcal{D}}^{\sigma}\}$, while the negative border is $\mathcal{B}d^{-}(\mathcal{F}_{\mathcal{D}}^{\sigma}) = \{X \in \mathcal{P}(\mathcal{I}) - \mathcal{F}_{\mathcal{D}}^{\sigma} \mid Y \subset X \text{ implies } Y \in \mathcal{F}_{\mathcal{D}}^{\sigma}\}$. Ideally, we can focus on the positive border of $\widetilde{\mathcal{F}}_{\mathcal{D}}^{\sigma}$, which consists of the set of maximal non-sensitive frequent itemsets, while hiding the minimal set of sensitive itemsets that comprise the negative border $\mathcal{B}d^{-}(\mathcal{S}_{\mathcal{D}}^{\sigma})$.

The constraint-based theory established in [18] depends on two constraints. The C_{sup} which is satisfied by itemsets having support greater than, or equal, to the support threshold σ and $C_{\overline{\text{sen}}}$ that is related to the non-sensitive itemsets. The conjunction of these constraints is adapted into an Apriori algorithm variation that aims to satisfy it and generates the positive and negative borders. This Algorithm performs a two-way pruning to remove candidates with infrequent and/or sensitive subsets and to remove sensitive itemsets. It is presented below in Algorithm 1.

4.3.2 Inverse Frequent Itemset Hiding Problem

We can now advance to the Inverse Frequent Itemset Hiding Problem and the database reconstruction approach that aims to solve it.

The Inverse Frequent Itemset Hiding Problem is defined as follows. Given a finite collection $\mathcal{F}_{\mathcal{D}}^{\sigma}$ of frequent itemsets X $\in \mathcal{F}_{\mathcal{D}}^{\sigma}$ and the supports of those items sup(X, $\mathcal{F}_{\mathcal{D}}^{\sigma}$), find a dataset D compatible with that set of collections and their frequencies [7].

A common approach associated to that problem is the Dataset Reconstruction. The algorithm followed in this paper is based in the one presented in [19] is cited in Algorithm 2. This is a naive top-down approach to database reconstruction. It starts with the itemset at the top of the lattice (largest k-itemset) and moves to lower levels (lowest being the literals). Then counts the cardinality of the top level itemsets and we can be certain that it is equal to their support, since they do not have any supersets. Then goes to the lower level (k–1) and subtracts from the cardinality of each set

Algorithm 1 Constraint-based Mining Algorithm

Input:
$\mathcal{D}$: transaction database
σ: the minimum support threshold
$\mathcal{B}$: Boolean formula representing $\mathcal{C}_{\overline{\text{sen}}}$
Output:
L: frequent itemsets satisfying $\mathcal{B}$
Description:
$C_1^{\mathcal{B}}$=candidate items that satisfy $\mathcal{B}$
L_1=the frequent items in $C_1^{\mathcal{B}}$
for (k=2; $L_{k-1} \neq \emptyset$; k++)
  $C_k^{\mathcal{B}}$=Apriori-genB (L_{k-1});
  for each transaction $t \in \mathcal{D}$
    C_t=**subset**($C_k^{\mathcal{B}}$, t)
    for each candidate $c \in C_t$
      $c.count$++;
  end
  $L_k = \{c \in C_k^{\mathcal{B}} | c.count \geq \sigma\}$
end
return $L = \bigcup_k L_k$

the total cardinality of its supersets. The resulting number (if positive) equals to the support of that specific itemset. Then proceeds accordingly to lower levels until the bottom. Based on the support found for each itemset, we create a database D' with that combination of itemsets and frequencies.

Algorithm 2 Database Reconstruction Algorithm

Input: P(I) with their support, k=length of largest itemset
Output: D' = reconstructed database
Description:
D'=∅, Q=∅, R=∅

While k>0:
  Q = {I_i || |I_i|=k && sup(I_i)>0}
  While Q ≠ ∅:
    p ⟵ remove(Q);
    Add itemset p in database D', sup(p) times
    Find all subsets of p: R = {f_i || f_i ⊂ p }
    Foreach q∈R:
      sup(q) = sup(q) - sup(p)
  k=k-1

4.4 Hiding Approach

Now, that we have defined both the Frequent Itemset Hiding Problem and its Inverse counterpart, alongside the constraint-based Border Theory, we will present our research on both of these fields and their combination. We start by describing the weakness that we found in the Database Reconstruction Algorithm [19]. We decided to use the same example as the one presented in their paper in order to make our point clear. So given the database in Table 4.1, the corresponding itemset lattice is in Fig. 4.1. The itemset lattice spans 3 levels, from the literal items A, B, C, D of the bottom level, to the second level with itemsets composed of the combinations of the literal items with length 2 and at the top similarly with length 3. In a general scenario a level k has itemsets that are created by combining k literal items.

The number in each level, next to the itemset, notes its support (the number of times that itemset is found, either the same or as a subset) in the transactional database. So for example if we search for A, it will be found 5 times in the original database 4.1. But if we take a look at the second level of the lattice and count how many {A}s there are, we will find 6. This happens because a number of those occurrences is combined in the next level. In detail, from supc(AC) = 2 and supc(AD) = 3, we get supc(ACD) = 2, so the total frequency of A, from those itemsets is 3, instead of 5.

Table 4.1 Sample database to demonstrate and compare our approach

Tid	Items
1	acd
2	c
3	acd
4	ab
5	cd
6	cd
7	bc
8	ad
9	bcd
10	a

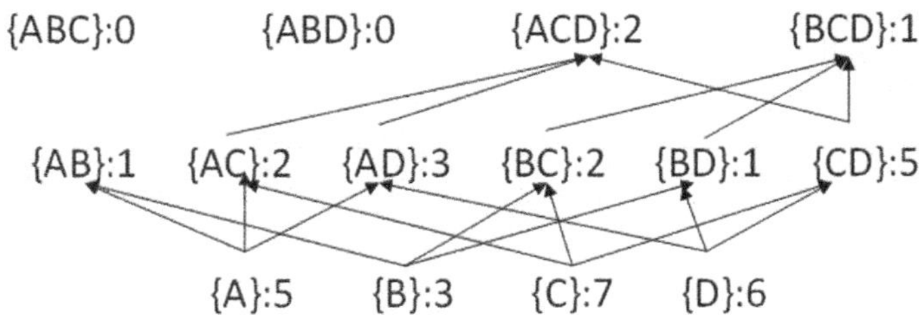

Fig. 4.1 Initial D2 lattice

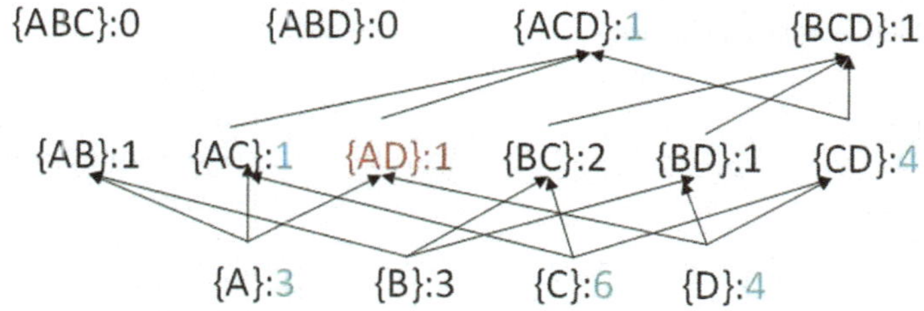

Fig. 4.2 Revised D2 lattice

Table 4.2 Reconstructed database after hiding

Tid	Items
1	acd
2	bcd
3	ab
4	bc
5	cd
6	cd
7	a
8	c

The above idea works as the foundation of the top-down (as seen in the itemset lattice) database reconstruction algorithm, but also presents some issues in its hiding aspects. So let's assume as in the original example that the sensitive itemset that we want to hide is {AD} and the support count is 2. According to the given algorithm, we decide to lower its support and in order to be hidden, we decrease by 2, so supc(AD) = 1. Now the itemset is hidden since its support is lowered by the threshold, but we have to adjust the database accordingly. The result is presented in Fig. 4.2. We highlight the sensitive itemset {AD} with red color and with light blue the changed support of the affected itemsets.

The algorithm, first, finds the subsets of {AD}, namely {A} and {D} and reduces their support by the same amount, 2. The idea is that, since the new support of {AD} is decreased by 2, then all of its subsets, {A} and {B}, will also appear two less times and need to be modified. Then it configures all the supersets of {AD}, now only {ACD}. It has been proven numerous times that the support of a superset can not be higher than its subsets and so, supc(ACD) = 1. Finally, similarly to the first idea the other subsets of {ACD} are found and their supports lowered by one. The reconstructed database given after the algorithm ends is given in Table 4.2.

We continue by presenting our findings on the above process. Initially, we examine the hiding efficiency and the data quality of the algorithm. In order to do that, we use the following metrics as in [20]. The Safety Measure $c_1 = \frac{(|S|-|\widetilde{S}|)}{|S|}$, where $|S|$ = cardinality of the frequent and sensitive itemsets in the database and $|\widetilde{S}|$ = the

Table 4.3 Metrics used in this paper

Name	Symbol	Explanation
Safety measure	C_1	Percentage of successfully hidden sensitive itemsets
Consistency measure	C_2	Percentage of non-sensitive itemsets maintained frequent
Data confidence	C_3	Percentage of true frequent itemsets in new database

cardinality of the set of sensitive itemsets that remain frequent in the newly reconstructed database. It represents the percentage of sensitive itemsets that have been successfully hidden (optimum $c_1 = 1$).

The Consistency Measure $c_2 = \frac{|\widetilde{\mathcal{N}}|}{|N|}$. $|N|$ = cardinality of the initial frequent itemsets in the database and $|\widetilde{\mathcal{N}}|$ = the cardinality of the set of frequent itemsets in the newly reconstructed database, without the sensitive itemsets. This measure gives the percentage of non-sensitive itemsets that have been maintained frequent (optimum $c_2 = 1$).

Finally, the Data Confidence $c_3 = \frac{(|\widetilde{\mathcal{R}}|-|F|)}{|\widetilde{\mathcal{R}}|}$. $|\widetilde{\mathcal{R}}|$ = cardinality of the total frequent itemsets in the new database and $|F|$ = the cardinality of the False Frequent itemsets, itemsets that were not frequent in the original database, but are frequent in the reconstructed one. This measure shows the percentage of proper frequent itemsets maintained after the transformation (optimum $c_3 = 1$) (Table 4.3).

So we proceed to use these metrics in the above example. The Safety Measure $c_1 = \frac{(|S|-|\widetilde{\mathcal{S}}|)}{|S|} = \frac{1-0}{1} = 1$, since {S}= {AD} and $\widetilde{\mathcal{S}} = \emptyset$, meaning that all sensitive itemsets that we want to hide are hidden in the new database. This is admittedly the most important measure since it shows if there are any privacy risks in the database that we may publish.

For the Consistency measure, we have {N}={AC,BC,CD,A,B,C,D} with $|N| = 7$ and $\widetilde{\mathcal{N}} = \{BC, CD, A, B, C, D\}$ with $|\widetilde{\mathcal{N}}| = 6$. As a result we get $c_2 = \frac{|\widetilde{\mathcal{N}}|}{|N|} = \frac{6}{7} = 0.86$. During the hiding process the support of {AC} is dropped below the threshold and as a result it is not frequent in the modified database. We lost an itemset that needs to be frequent and this measure depicts in this case that the data quality of the new database is not the same to the original.

For the Data Confidence, by selecting to decrease the supports of various itemsets in our hiding methodology, it will always result in $c_3 = \frac{(|\widetilde{\mathcal{R}}|-|F|)}{|\widetilde{\mathcal{R}}|} = 1$, because no False Frequent Itemsets will be produced in the hiding process. If instead we chose to increase the support thresholds to hide sensitive knowledge, we should compute and evaluate this metric properly.

Overall the results of the proposed methodology in [19], appear promising in that sample. The sensitive data is hidden, no falsely frequent itemsets are presented and the data quality is acceptable. Further testing is always necessary for any algorithm, so we decided to prepare two other scenarios and we present our results here. For the

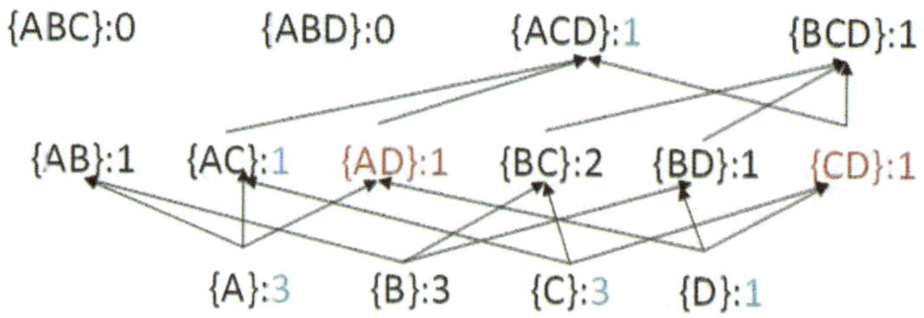

Fig. 4.3 Revised lattice for the second experiment

first one, we chose exactly the same dataset as in Table 4.1 and the same threshold, so we also have the same initial lattice as in Fig. 4.1, but for the hiding process we chose a different sensitive set.

The sensitive set, for our second experiment, consists of S = {AD,CD}. We follow the same steps. First, same as before we decrease the support of supc(AD) = 1 and modify accordingly the supports of its superset {ACD}, its subsets {A} and {D} and its superset's subsets {AC}, {CD} and {C}. Then we do the same for {Cd}, with its support decreased by 3 to supc(CD) = 1. Its supersets {ACD} and {BCD}, both have support supc(ACD) = supc(BCD) = 1≤supc(CD), so no need for adjustments there. We still need to decrease the support of its subsets {C} and {D}, which results in supc(C) = 3 and supc(D) = 1 and the lattice in Fig. 4.3. A problem presents here, in the Database Reconstruction algorithm of this lattice. f(D) = supc(D) – f(ACD) – f(BCD) = 1 – 1 – 1 = –1 which leads to a failure. This, in turn means that the frequencies of the produced lattice can not be the outcome of a frequent itemset mining process (like the Apriori algorithm) on a real dataset and the the reconstruction fails.

To counter this issue, we changed the algorithm a bit. The basic idea exists in the original, but we adjusted it for hiding purposes (instead of just database reconstruction). For any f(X)≤ −1, where X⊂S, a proper subset of a sensitive set that we try to hide, we will increase its support to the point where f(X) = 0. This will not result in further issues (such as False Positives), because in the initial database all subsets of a sensitive and frequent itemset is also frequent. Thus, the increase in support of these subsets can only result in them being frequent in the new database, when they would not be (even if the reconstruction algorithm worked). Back to our example, this means that we will increase the supc(D) = 2, so f(D) = 0. This allows us to reconstruct the database in Table 4.4 and also makes {D} a frequent itemset, as seen in the new lattice in Fig. 4.4.

We also check the metrics against this new reconstructed database to evaluate this approach. For the Safety Measure, we have {S}= {AD,CD} and $\widetilde{S} = \emptyset$, so $c_1 = 1$, we achieved total hiding. For Consistency measure, we have {N}={AC,BC,A,B,C,D} with $|N| = 6$ and $\widetilde{\mathcal{N}} = \{BC, A, B, C, D\}$ with $|\widetilde{\mathcal{N}}| = 5$, so $c_2 = 0.83$, pretty close to the previous one. Again we have no False Positives so $c_3 = 1$. The adjusted algorithm performs the way it was expected to and similar to the original.

We proceed to our final approach. **We wanted to combine the constraint-based approach with the database reconstruction algorithm**. The algorithm that we

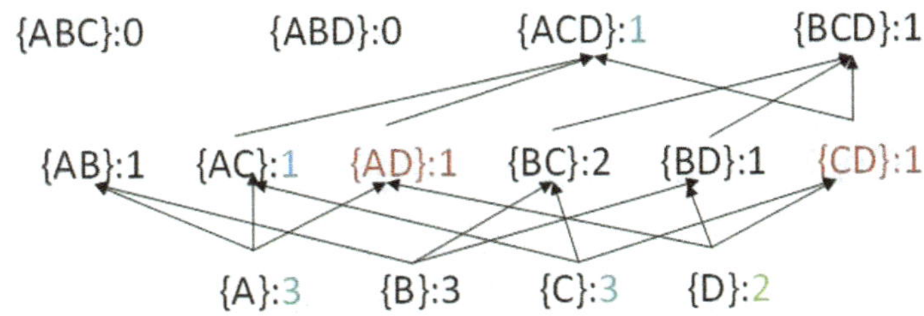

Fig. 4.4 Proper revised lattice for the second experiment after adjustments

Table 4.4 Reconstructed database for adjusted problem and algorithm

Tid	Items
1	acd
2	bcd
3	ab
4	bc
5	a

developed to accomplish that is presented in Algorithm 3. We need as input the transactional database D, the minimum support threshold σ and the set of sensitive itemsets S that the user wants to hide. We start by applying the AprioriGenB first. This is the Apriori variation mentioned earlier that performs a two way pruning, finding and separating the sensitive itemsets in the usual Apriori sequence, creating the Positive and Negative Borders and their lattice in the process. We then select the constraint-based inverse itemset lattice with reduction of supports as our hiding approach. We give as input to the database reconstruction algorithm, not the whole initial lattice but the pruned one by our Apriori, meaning the lattice produced at the previous step. The algorithm adjusts the frequencies of the itemsets in the Negative Border(and its subsets) in order to properly hide the sensitive data.

By doing this we provide to the database reconstruction the maximal set of frequent and non-sensitive itemsets (Positive Border) and the minimal set of sensitive itemsets that need to be hidden (Negative Border) and we assure that no sensitive itemsets will be in the new database. It is highly likely with this change, that some itemsets will have f(X)≤ -1. According to the idea provided earlier, we can increase the supports of subsets of the positive border **without sacrificing data quality or data security**, since these itemsets will, by definition, also be frequent and non-sensitive. For the sets of the Negative Border, we can also safely decrease their support since our goal is to hide them and as for their subsets we can increase it, similarly to those of the positive border, if it is needed by the reconstruction algorithm, since all the sets in the Negative Borders are frequent, their subsets will be frequent as well.

Showcasing this to our example the Border Based lattice of the sample database with S = {AD} is given in Fig. 4.5. The Positive Border that we get from the initial dataset is $\mathcal{B}d^{+}$ = {BC:2,AC:2,CD:5} and the Negative Border is $\mathcal{B}d^{-}$ = {AD:3}.

This means that all 3-length itemsets are omitted and the same applies to {AB} and {BD}. Now, in order to hide {AD} (our sensitive itemset) we decrease its support by 2, so sup(AD) = 1 and in turn we adjusted the affected sup(A) = sup(A) − 2 = 3. The support of itemset {D} would also be decreased, but it will increase in the next step, so in the end it remains unchanged. Finally, we need to increase the support of {C} by 2 to properly reconstruct the database which corresponds to Table 4.5 and the lattice in Fig. 4.6.

Applying our metrics to our last example, we can easily see that the Safety Measure is $c_1 = 1$, since $\widetilde{\mathcal{S}} = \emptyset$, **we have successfully hidden all sensitive data**. The Data Confidence is also $c_3 = 1$, **because our algorithm does not create False Frequent itemsets**. Finally, the Consistency measure $c_2 = 1$, we have perfect Data Quality in the new database. The initial $|N| = 7$ and $\widetilde{\mathcal{N}} = \{AC, BC, CD, A, B, C, D\}$ with $|\widetilde{\mathcal{N}}| = 7$. **This means that our adjusted algorithm outperforms the original algorithm in Data Quality in the example dataset the original paper provided.**

Algorithm 3 Border Base Reconstruction Algorithm

Step 1: Get input
$\mathcal{D}$: transaction database
σ: the minimum support threshold
S: set of sensitive itemsets that need to be hidden
Step 2: Constraint based AprioriGenB
We apply our AprioriGenB algorithm to get the pruned frequent itemsets and create the border-based itemset lattice.
Step 3: Select Hiding Approach
We select the constraint based Inverse Itemset Lattice and give it as input data the sets resulted by the border revision of our Apriori.
Step 4: Adjust Lattice
We adjust the support of the itemsets affected by the hiding process for the data reconstruction. We can safely increase the support of any itemset of the positive border since they are all frequent and non-sensitive and decrease the supports of the itemsets of the negative border.
Step 5: Data Reconstruction
Having adjusted the frequencies of all itemsets of the lattice in the previous step, in such a way that the data reconstruction algorithm does not fail, we create the new sanitized database .
Output: Reconstructed database D' that is safe to publish.

4.5 Conclusion and Future Steps

In every problem, the study of its inverse is equally important to the original problem. The same should apply to the Frequent Itemset Mining Problem. Few studies have been conducted on its reverse version, in comparison to the vast and thorough research of the initial problem. The main goal for studying both, in the aspect of privacy preserving data mining, is the optimization of knowledge hiding in terms of volume-space, computation complexity in resources and time. We need to create a

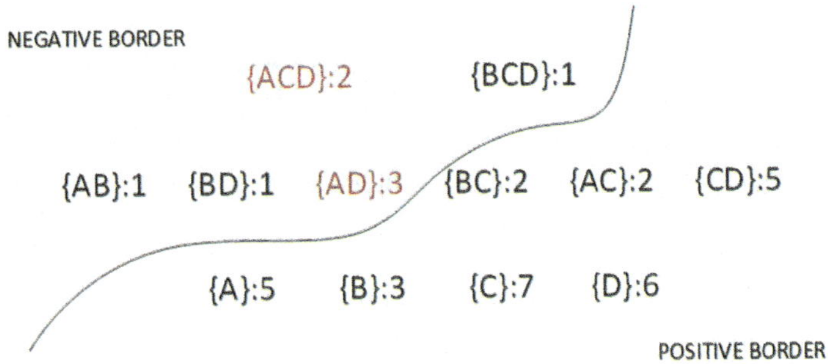

Fig. 4.5 Border based approach lattice

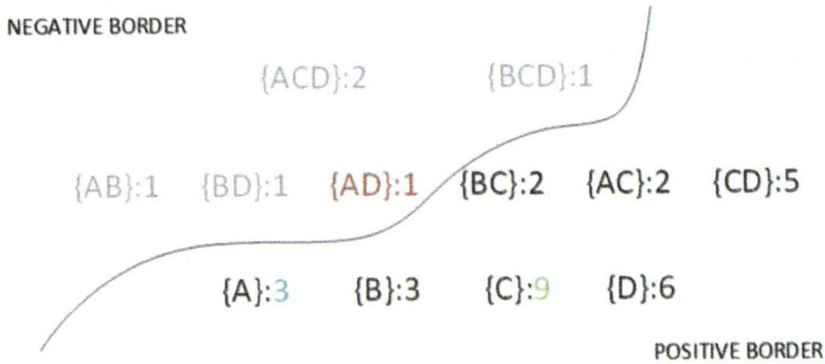

Fig. 4.6 Border based revised lattice

Table 4.5 Reconstructed database for border based approach

Tid	Items
1	bc
2	bc
3	ac
4	ac
5	cd
6	cd
7	cd
8	cd
9	cd
10	ad
11	b

transformed database that the sensitive itemsets become hidden, while the derivable knowledge is maintained as much as possible. In this paper, we present a database reconstruction approach for the hiding of sensitive knowledge employing a constraint based Apriori algorithm. It is a novel approach to resolve Inverse Frequent Itemset Mining Problem that keeps data quality of the initial dataset to the sanitized one after the database reconstruction. The second advantage to previous approaches that use additional records to perform hiding is that the produced sanitized result maintains the lowest possible footprint in terms of volume-space.

Future steps include, making extensive experiments to assess further the efficiency of our proposed approach, both in terms of computational costs and in terms of data quality. The direction of the parallelization of database reconstruction methods should also be considered, since it is a field that has not yet been researched and could produce noteworthy results.

Acknowledgements The authors of this paper would like to thank the University of Piraeus Research Center for the financial support of this research paper.

References

1. C. Clifton, D. Marks, Security and privacy implications of data mining, in *SIGMOD*, pp. 15–19 (1996)
2. R. Agrawal, R. Srikant, Privacy-preserving data mining, in *SIGMOD*, pp. 439–450 (2000)
3. L. Caruccio, D. Desiato, G. Polese, G. Tortora, Gdpr compliant information confidentiality preservation in big data processing, in *IEEE Access*, pp. 205034–205050 (2020)
4. V.S. Verykios, Association rule hiding methods. Wiley Interdiscip. Rev. Data Mining Knowl. Discov. **3**, 28–36 (2013)
5. M. Cafaro, M. Pulimeno, Frequent itemset mining, in *Business and Consumer Analytics: New Ideas*, pp. 269–304 (2019)
6. P. Krasadakis, V.S. Verykios, E. Sakkopoulos, Parallel based hiding of sensitive knowledge, in *ICTAI* (2020)
7. T. Mielikäinen, On inverse frequent set mining problems, in *Proceedings of the 2nd Workshop on Privacy Preserving Data Mining*, pp. 18–33 (2003)
8. M. Atallah, E. Bertino, A. Elmagarmid, M. Ibrahim, V. Verykios, Disclosure limitation of sensitive rules, in *KDEX Workshop*, pp. 45–52 (1999)
9. V. Verykios, A. Elmagarmid, E. Bertino, Y. Saygin, E. Dasseni, Association rule hiding, in *KDE*, pp. 434–447 (2004)
10. A. Gkoulalas-Divanis, V. Verykios, Hiding sensitive knowledge without side effects. KIS **20**, 263–299 (2009)
11. E.C. Stavropoulos, V.S. Verykios, V. Kagklis, A transversal hypergraph approach for the frequent itemset hiding problem. Knowl. Inf. Syst. 625–645 (2016)
12. Y. Wang, X. Wu, Approximate inverse frequent itemset mining: Privacy, complexity, and approximation, in *ICDM'05* (2005)
13. Y. Guo, Reconstruction-based association rule hiding, in *SIGMOD2007* (2007)
14. A. Guzzo, L. Moccia, D. Sacca, E. Serra, Solving inverse frequent itemset mining with infrequency constraints via large-scale linear programs. ACM TKDD **7** (2013)
15. L. Shaoxin, M. Nankun, L. Junqing, L. Xiaofeng, Privacy preserving frequent itemset mining: Maximizing data utility based on database reconstruction. Comput. Secur. **84**, 17–34 (2019)

16. D. Sacca, E. Serra, A. Rullo, Extending inverse frequent itemsets mining to generate realistic datasets: complexity, accuracy and emerging applications. Data Min. Knowl. Disc. **13**, 1736–1774 (2019). (Nov.)
17. F. Gao, A. Khandelwal, J. Liu, Mining frequent itemsets using improved apriori on spark, in *ICISDM*, pp. 87–91 (2019)
18. V. Verykios, E. Stavropoulos, V. Zorkadis, A. Elmagarmid, A constraint-based model for the frequent itemset hiding problem, *e-Democracy*, pp. 49–64 (2019)
19. X. Chen, M. Orlowska, X. Li, A new framework of privacy preserving data sharing, in *Computer Science* (2004)
20. C. Makris, P. Markovits, Evaluation of sensitive data hiding techniques for transaction databases, in *SETN '18*, pp. 1–8 (2018)

Chapter 5
A Rough Inference Software System for Computer-Assisted Reasoning

Zbigniew Kokosiński and Kacper Jaworski

Abstract In this chapter construction and implementation of a new rough inference software system is described. This artificial intelligence tool is designed for information systems for building and processing of decision tables, reduction of attributes and generation of nonredundant rules that support the decision process. It is based on new approximate algorithms developed for processing multivalued decision tables and implemented in MATLAB. One software application in electrical engineering is presented as an illustrative example.

Keywords Rough inference system · Information system · Decision table · Approximate algorithms · DGA · Power transformer diagnostics

5.1 Introduction

In this chapter a new rough inference software system is described. The system was developed as an artificial intelligence tool in all application areas where the rough set theory provides an adequate mathematical framework for modeling and decision making [13, 18]. The key features of such systems include building and processing of decision tables, reduction of decision table size by removing redundant attributes and generation of nonredundant rules that support the decision process in an information system.

Decision tables are built of records which fields are multivalued attributes. The attributes represent features of the modeled system. All possible combinations of features define state space of the decision system.

Proper modeling of a system at hand require adequate handling of nonredundant state space, providing that different classes of admissible states and inadmissible states are distinguishable. The redundant attributes can be removed from the decision table [2, 14, 19, 23, 24].

Z. Kokosiński (✉) · K. Jaworski
Faculty of Electrical & Computer Engineering, Cracow University of Technology, ul. Warszawska 24, 31-155 Kraków, Poland
e-mail: zk@pk.edu.pl

M. Virvou et al. (eds.), *Advances in Artificial Intelligence-based Technologies*, Learning and Analytics in Intelligent Systems 22,
https://doi.org/10.1007/978-3-030-80571-5_5

System attributes can be mutually related or unrelated. None of the attributes may be removed out of the nonredundant set of unrelated attributes because at least one pair of system states that belong to different classes is no longer differentiated. From that point of view searching for a nonredundant set of attributes leads in a natural way to searching for a minimal attribute set and the search criterion is purely quantitative. On the other hand, when system attributes are mutually related the problem of attribute reduction transforms to a feature selection problem which nature is qualitative [1, 3, 7, 12, 15, 17]. The feature selection is a hot topic in contemporary research. Basic techniques applied for solving this problem are various wrapper and filter methods and their combinations [17, 22].

It is well known that combinatorial problems to be solved in implementing any nontrivial rough inference system belong to the NP complexity class [5, 10]. Therefore, approximation algorithms have to be used.

In this chapter a new approximate algorithm for multivalued attribute reduction in information systems is presented. The second approximate algorithm that appears in two basic variants is devoted to nonredundant rule generation in decision tables. Boths variants were inspired by methods of combinational circuit synthesis developed by Kapralski [9] and Kozakov [11]. The logical circuits are represented there by Boolean matrices and can be interpreted as the simplest information systems.

The developed software that supports the decision process was developed on the basis of new algorithms for processing multivalued decision tables and implemented in MATLAB. It is able to read, generate and process decision tables in order to compute nonredundant solutions in the form of a rule set.

The chapter is self-contained and consist of the theoretical background concerning rough sets, pseudocodes of algorithms, several computational examples, and a description of the developed software tool. An application in electrical engineering is presented as an illustrative example.

5.2 Basic Concepts

In this section, we will recall basic concepts of the rough set, the information system, and the decision system.

5.2.1 Rough Sets

The theory of rough sets was introduced by Zdzisław Pawlak and provides mathematical tools for describing incomplete and inexact information to be processed in information systems [18].

Objects o of an universum U having identical attributes are called similar and are used for creating elementary sets. Sum of any elementary sets is called a defined or exact set. Undefined sets are called rough sets and they can not be characterized in a unique way by attributes of their elements. Therefore, every rough set can be characterized by two defined sets being its lower and upper approximations. The difference of the upper and lower approximation sets is called the boundary region. Lower approximation consists of elements that definitely belong to the given set while the upper approximation includes in addition all elements which - on the basis of the existing knowledge - can not be excluded from that set. The set is a rough set if and only if its boundary region is not empty. The theory of rough sets enable many useful operations [18]:

- search for data dependencies,
- reduction of input data size,
- determining of input data weights,
- generation of decision rules from input data.

The theory of rough sets provides many advantages for the information system designer:

- there is no need for data to be characterized by probability or fuzziness,
- there exist algorithms for analysis of data with rough characteristics,
- interpretation of results is straightforward,
- the computations involve basic mathematics, only.

In this chapter, we deal with rough sets applications to decision support systems.

5.2.2 *Information System*

Let us define the information system IS:

$$IS =< U, A, V, f >$$

where:
U—is a finite, non empty set of objects (the universe): $U = \{o_1, o_2, \ldots, o_n\}$;
A—is a finite, non empty set of attributes: $A = \{A_1, \ldots, A_j, \ldots, A_m\}$;
V_j—is a finite, non empty set of attribute values in A_j, $j \in \{1, \ldots, m\}$:
$V_j = \{a_1, a_2, \ldots, a_{\alpha_j}\}$, where $\alpha_j = |V_j|$; V_j is called the domain of the attribute $A_j \in A$;
V—is a finite set of attribute values in all $V_j : V = \cup V_j$;
f—is an information function, which corresponds to a mapping of Cartesian product $U \times A \rightarrow V : \forall_{o_i \in U}, i \in \{1, \ldots, n\}, A_j \in A : f(o_i, A_j) \in V_j$.

IS is a tabular data structure that stores essential data of the system. Columns of IS table are usually labeled by attributes (parameters, properties, features) while rows correspond to objects (situations, states). Table rows contain combinations of attributes' values for all distinguishable states.

Example 1 An example of information system IS defined below is shown in two representations in Tables 5.1 and 5.2, where: $U = \{o_1, \ldots, o_5\}$, $A = \{A_1, \ldots, A_4\}$, $V_1 = \{1, 2\}$ and $\alpha_1 = |V_1| = 2$, $V_2 = \{3, 4, 5\}$ and $\alpha_2 = |V_2| = 3$, $V_3 = \{6, 7, 8\}$ and $\alpha_3 = |V_3| = 3$, $V_4 = \{9, 10, 11, 12\}$ and $\alpha_4 = |V_4| = 4$.

Without loss of generality in our algorithms nonrelated attribute domains V_j may be normalized as follows: $V_j = \{a_1, a_2, \ldots, a_{\alpha_j}\} = V_j = \{1, 2, \ldots, \alpha_j\}$. □

5.2.3 Decision System

The basic difference between decision system DS and IS relies on partitioning the set of attributes A into disjoint subsets C and D. DS is defined as follows:

$$DS =< U, C, D, V, f >$$

where:
U and V satisfy the same definitions as in IS;
$C, D \subset A$; $C, D \neq \emptyset$; $C \cup D = A$; $C \cap D = \emptyset$—the elements of C are called conditional attributes while the elements of D are called decision attributes;

Table 5.1 IS table for Example 1

Objects U	Attributes A			
	A_1	A_2	A_3	A_4
o_1	a_1	a_3	a_6	a_9
o_2	a_2	a_3	a_7	a_{10}
o_3	a_1	a_4	a_8	a_9
o_4	a_2	a_4	a_6	a_{11}
o_5	a_2	a_5	a_6	a_{12}
$\alpha_j =$	2	3	3	4

Table 5.2 Normalized IS table for Example 1

Objects U	Attributes A			
	A_1	A_2	A_3	A_4
o_1	1	1	1	1
o_2	2	1	2	2
o_3	1	2	3	1
o_4	2	2	1	3
o_5	2	3	1	4
$\alpha_j =$	2	3	3	4

Table 5.3 DS table for Example 2

Objects U	Conditions C			Decisions D
	C_1	C_2	C_3	D_1
o_1	a_1	a_3	a_6	a_9
o_2	a_2	a_3	a_7	a_{10}
o_3	a_1	a_4	a_8	a_9
o_4	a_2	a_4	a_6	a_{11}
o_5	a_2	a_5	a_6	a_{12}
$\alpha_j =$	2	3	3	4

Table 5.4 Normalized DS table for Example 2

Objects U	Conditions C			Decisions D
	C_1	C_2	C_3	D_1
o_1	1	1	1	1
o_2	2	1	2	2
o_3	1	2	3	1
o_4	2	2	1	3
o_5	2	3	1	4
$\alpha_j =$	2	3	3	4

f—is a decision function, which corresponds to a mapping of Carthesian product $U \times C \rightarrow D : \forall_{o_i \in U}, i \in \{1, \ldots, n\}, C_j \in C : f(o_i, C_j) \in D_k, \;\; j \in \{1, \ldots, |C|\}$, $k \in \{|C| + 1, \ldots, m\}$.

Moreover, identical values of $a_i \in D_k$, $i \in \{1, \ldots, \alpha_j\}$, define decision classes (equivalency relations) DC composed of objects o_i or their associated combinations of attributes $a_j \in C$.

Example 2 An example of decision system DS defined on the basis of IS from Table 5.1 is shown in two representations in Tables 5.3 and 5.4, where $U = \{o_1, \ldots, o_5\}, C = \{C_1 = A_1, C_2 = A_2, C_3 = A_3\}, D = \{D_1 = A_4\}, V_1 = \{1, 2\}$ and $\alpha_1 = |V_1| = 2$, $V_2 = \{3, 4, 5\}$ and $\alpha_2 = |V_2| = 3$, $V_3 = \{6, 7, 8\}$ and $\alpha_3 = |V_3| = 3$, $V_4 = \{9, 10, 11, 12\}$ and $\alpha_4 = |V_4| = 4$. $DC = \{o_1, o_3\}\{o_2\}\{o_4\}\{o_5\}$.

As in the case of DS, attribute domains V_j may be normalized as follows: $V_j = \{a_1, a_2, \ldots, a_{\alpha_j}\} = V_j = \{1, 2, \ldots, \alpha_j\}$. □

5.2.4 Indiscernibility Relation

Rough set theory models similarities and differences between objects based on the notions of indiscernibility and discernibility [21, 25]. Indiscernibility relation in IS exists iff there are objects $o_i \in U$ that have identical attributes subsets. We call such objects indiscernible with respect to attributes. For instance, o_1 and o_3 in Table 5.1 are indiscernible with respect to attributes A_1 and A_4.

The indiscernibility relation can be defined as follows:

$$IS =< U, B, V, f > \text{and} B \subseteq A$$

$$IND_{IS}(B) = (o_g, o_h) \in U \times U : \forall A_j \in B : f(o_g, A_j(g)) = f(o_h, A_j(h))$$

where:
(o_g, o_h)—is a pair of objects that belongs to Cartesian product $U \times U$;
A_j - is any attribute in B;
$f(o_g, A_j(g))$—is an information function for a pair $(o_g, A_j(g))$.

Pairs of objects belong to the relation if they possess identical values for all attributes in B. Indiscernibility relation $IND_{IS}(B)$ is at the same time an equivalence relation (partitioning) which is reflexive, symmetrical and transitive.

5.3 The Approximate Algorithms for Information Systems

It is well known that combinatorial problems to be solved in implementing any non-trivial rough inference system belong to the NP complexity class, therefore approximation algorithms have to be used [5]. We developed new approximate algorithms for attribute reduction in decision tables and for generation of optimal sets of rules. The new algorithms and their variants are used for processing multivalued decision tables. They are generalizations of approximate algorithms for solving NP-complete BASE problem defined in Boolean matrices [10] as well as approximate algorithms for solving NP-complete MINIMUM DISJUNCTIVE NORMAL FORM (MDNF) problem [4, 5] (problem LO9 in [5]). In opposition to the above sources, we apply consequently the structural programming paradigm.

Two main algorithms used in our program are: MV-DRMAX and DT-RULE-GEN. They are described in detail in the next subsections.

5.3.1 The Approximate Algorithm for Attribute Reduction

Approximate algorithms developed for solving BASE problem in Boolean tables [10] are useful for computing a minimal reduct in two-valued decision tables. They can be generalized for multivalued matrices and their reducts.

In this subsection we will focus on efficient reduction of information tables built of redundant, unrelated, multivalued attributes. In constructing our algorithm we were inspired by the approximate algorithm DRMAX [10], which can be used: 1. for argument reduction in binary input matrices (in logical synthesis of combinational circuits), and 2. for partitioning binary output matrices (when a decompositions of combinational logic is needed).

Let:
$A[n, m]$—be an input matrix,
B—be a minimal or approximate reduct of A, $B \subseteq A$,
$\alpha(j)$—denotes the number of different values of attributes in A_j columns,
$max(a_j)$ and $min(a_j)$—denote the maximal and minimal number of a_i occurences in A_j, respectively,
$\beta(j)$—denotes in A_j the absolute value of difference $max(a_j) - min(a_j)$,
$\rho_A = \rho(A)$—denotes the number of different rows in A.

The approximate algorithm MV-DRMAX (Multivalued Different Row MAXimum) for attribute reduction in redundant decision tables can be formulated as follows:

Algorithm MV-DRMAX

1. Reduce input table A by deleting multiple occurences of rows and columns, next delete single-valued columns of A, if they exist.
2. Create the empty base $B = \emptyset$.
3. Select jth column from table A for which function $\beta(j)$ is minimum and function $\alpha(j)$ is maximum. In the case of ties draw jth column at random.
4. Compute the set sum $B = B \cup A_j$ and reduce $A = A \setminus A_j$.
5. Repeat

   5.1. Find the next column $A_j \in A$, such that $\rho(B \cup A_j)$ is maximum (in the case of ties draw A_j at random).
   5.2. Update $B = B \cup A_j$ and $A = A \setminus A_j$.

   Until $(\rho(B) = \rho_A)$.
6. While (there exist $B_j \in B$ such that $\rho(B \setminus B_j) = \rho_A$) Do

   6.1 $B = B \setminus B_j$. (final attribute reduction)
7. Return base B (eventually extended by rows deleted from A in step 1).

Example 3 Find a minimum base B of a normalized table A (Table 5.5) using algorithm MV-DRMAX. Table A contains neither redundant rows nor columns.
$\rho_A = \rho(A) = 10$.
$B = \emptyset$.
$\beta(3) = \min$ and $\alpha(3) = \max$.
$B = B \cup \{A_3\}$.
$\rho(A_1, A_3) = 6$, $\rho(A_2, A_3) = 6$, $\rho(A_3, A_4) = 8 = \max$, $\rho(A_3, A_5) = 5$,

Table 5.5 Normalized A table for Example 3

Objects U	Attributes A					
	A_1	A_2	A_3	A_4	A_5	A_6
o_1	1	1	1	1	1	1
o_2	1	2	2	2	1	1
o_3	1	3	2	1	2	2
o_4	1	3	3	1	2	2
o_5	1	3	3	1	1	1
o_6	2	2	3	2	2	1
o_7	2	2	3	3	1	1
o_8	2	3	2	1	2	2
o_9	3	2	1	3	1	1
o_{10}	3	2	1	2	1	2
$\alpha(j)$	$3(max)$	$3(max)$	$3(max)$	$3(max)$	2	2
$max(a_j)$	5	5	4	5	6	6
$min(a_j)$	2	1	3	2	4	4
$\beta(j)$	3	4	$1(min)$	3	2	2

$\rho(A_3, A_6) = 6$.
$B = \{A_3, A_4\}$.
$\rho(A_1, A_3, A_4) = 9 = \max$, $\rho(A_2, A_3, A_4) = 8$, $\rho(A_3, A_4, A_5) = 9 = \max$,
$\rho(A_3, A_4, A_6) = 9 = \max$.
$B = \{A_1, A_3, A_4\}$.
$\rho(A_1, A_2, A_3, A_4) = 9$, $\rho(A_1, A_3, A_4, A_5) = \rho(A_1, A_3, A_4, A_6) = 10 = \rho_A$.
$B = \{A_1, A_3, A_4, A_5\}$.
Thus, $B = \{A_1, A_3, A_4, A_5\}$ (or $B = \{A_1, A_3, A_4, A_6\}$) and $|B| = 4$. □

5.3.2 The Algorithm for Approximate Rule Generation

For the purpose of generation of nonredundant rules in a rough inference system we developed two versions of the algorithm inspired by methods of searching minimal disjunctive form of combinational dependence of a Boolean vector y with Boolean matrix X [9, 11]. The formulation of minimization problem in [9] was different from minimization of Boolean function problem since presence of Don't Care rows in X was allowed (logic functions are fully determined). Our aim is to generalize that approach to multivalued tables and present both algorithms according to structural programming paradigm.

DT-RULE-GEN (Decision Table - RULE GENeration) is an approximate algorithm for generation of nonredundant rule set for well-determined decision table.

DV-RULE-GEN (Decision Vector - RULE GENeration) is an approximate algorithm for generation of nonredundant rule set for well-determined decision table with single decision vector. DV-RULE-GEN appears in the two versions DV-RULE-GEN-A and DV-RULE-GEN-B. DT-RULE-GEN applies one of them.

Let:
DT—is a normalized decision table, $DT = C \cup D$,
$C[n, m]$—is a multivalued input table of n input attributes,
$D[n, p]$—is a multivalued table of p decision attributes,
D_j—is a jth decision vector in D, where $j \in \{1, \ldots, p\}$,
$\alpha(j)$—denote the number of different values of attributes in D_j columns ,
D^i_j—is an ith decision vector for decomposed D_j, where $i \in \{1, \ldots, \alpha(i)\}$;
(for D^i_j in our examples we use a simplified notation D_{ji}).
RL^i_j—is an ith list of rules in the sum-of-product form generated for $D^i_j(q) = i$, $q \in \{1, \ldots, n\}$,
CL^i_j—is an ith list of numbers of all $D^i_j(q) = i$ represented by rules in RL^i_j,
RL, CL—are lists of lists of rule sets and cover sets, correspondingly.

Algorithm DT-RULE-GEN

1. Check if DT is well-determined. If not then EXIT.
2. Setup empty RL, CL.
3. For $j = 1$ to p Do
   - 3.1. For D_j create $\alpha(j)$ binary decision tables $D^i_j = 0$, $1 \le i \le \alpha(j)$.
   - 3.2. For $t = 1$ to n Do If $(D_j(t) = i$ Then $D^t_j(t) = i$.
   - 3.3. For $i = 1$ to $\alpha(j)$ Do
     - 3.3.1. **DV-RULE-GEN**$(C, D^i_j, j, RL^i_j, CL^i_j)$
     - 3.3.2. $RL = RL \cup RL^i_j)$
     - 3.3.3. $CL = CL \cup CL^i_j)$
4. **DT-MINSETCOVER** (RL, CL)
5. **DT-RULES-OUTPUT** (RL)

DT-MINSETCOVER(RL, CL) - computes the minimal lists RL of nonredundant rules generated for DT.
DT-RULES-OUTPUT(RL) - outputs RL.

Details of DT-MINSETCOVER (RL, CL) and DT-RULES-OUTPUT (RL) will be omitted here.

Table 5.6 *DT* table for Example 4

Objects	C_1	C_3	C_4	C_5	D_1	D_{11}	D_{12}	D_{13}
o_1	1	1	1	1	1	1	0	0
o_2	1	2	2	1	2	0	2	0
o_3	1	2	1	2	3	0	0	3
o_4	1	3	1	2	3	0	0	3
o_5	1	3	1	1	1	1	0	0
o_6	2	3	2	2	1	1	0	0
o_7	2	3	3	1	1	1	0	0
o_8	2	2	1	2	3	0	0	3
o_9	3	1	3	1	2	0	2	0
o_{10}	3	1	2	1	2	0	2	0

Generated rules:

1. $C_1(1)$ AND $C_3(1) \to D_{11}$,
2. $C_1(2)$ AND $C_3(3) \to D_{11}$,
3. $C_3(3)$ AND $C_5(1) \to D_{11}$,
4. $C_1(3) \to D_{12}$,
5. $C_1(1)$ AND $C_4(2) \to D_{12}$,
6. $C_1(1)$ AND $C_4(2) \to D_{12}$,
7. $C_1(1)$ AND $C_5(2) \to D_{13}$.

Computation time: 30.55 ms.

Algorithm DV-RULE-GEN-A$(C, D_j^i, j, RL_j^i, CL_j^i)$

1. Repeat

   1.1. For the first uncovered $D_j^i(q) = i$ Do
   1.2.1. Set $k = 1$.
   1.2.2. Repeat
   1.2.2.1. Generate in a linear order consecutive k-subset C' of C.
   1.2.2.2. If (there exists a rule in C' that covers $D_j^i(q) = i$
   different from all rules for $D_j^i(r) = 0$ Then
   add it to RL_j^i and update CL_j^i for all $D_j^i(s) = i$
   with identical rule in C'); $q, r, s \in \{1, \ldots, n\}$.
   1.2.2.3 If all C(p,k) k-subsets have been generated Then
   If $k < p$ Then $k = k + 1$.
   Until $D_j^i(q) = i$ is covered in C'.

   Until all $D_j^i(q) \neq 0$ are covered.

Example 4 Generate rules for a normalized and reduced table *DT* depicted in Table 5.6 using algorithm DV-RULEGEN-A. Columns C_1, C_3, C_4, C_5 correspond one-to-one to the base columns B_1-B_4. For D_j^i we use a simplified notation D_{ji}. Order of rule generation is preserved. □

DV-RULE-GEN-B is another approximate algorithm for generation of rule set for any well-defined decision table. From the previous algorithm it differs with a way of checking existance of the searched rule. In opposition to DV-RULE-GEN-A the algorithm DV-RULE-GEN-B checks in B' if there exist rules for all decision vectors D_i^j in the same step, not separately one by one.

The algorithm DV-RULE-GEN-B runs as follows:

Algorithm DV-RULE-GEN-B$(C, D_j^i, j, RL_j^i, CL_j^i)$

1. Set $k = 1$.
2. Repeat

   2.1. Repeat

   2.1.1. Generate in a linear order consecutive k-subset C' of C.

   2.1.2. For every $D_j^i(q) = i$ uncovered in C' Do

   2.1.2.1 If (there exists a rule in C' that covers $D_j^i(q) = i$, different from all rules for $D_j^i(r) = 0$) Then add it to RL_j^i and update CL_j^i for all $D_j^i(s) = i$ with identical rule in C'); $q, r, s \in \{1, \ldots, n\}$.

   Until all C(p,k) k-subsets is generated.

   2.2. If $k < p$ then $k = k + 1$.

   Until all $D_j^i(q) \neq 0$ are covered.

Example 5 Generated rules for a normalized and reduced table DT depicted in Table 5.7 using algorithm DV-RULEGEN-B. olumns C_1, C_3, C_4, C_5 correspond one-to-one to the base columns B_1-B_4. For D_j^i we use a simplified notation D_{ji}. Order of rule generation is preserved. □

Table 5.7 DT table for Example 5

Objects	C_1	C_3	C_4	C_5	D_1	D_{11}	D_{12}	D_{13}
o_1	1	1	1	1	1	1	0	0
o_2	1	2	2	1	2	0	2	0
o_3	1	2	1	2	3	0	0	3
o_4	1	3	1	2	3	0	0	3
o_5	1	3	1	1	1	1	0	0
o_6	2	3	2	2	1	1	0	0
o_7	2	3	3	1	1	1	0	0
o_8	2	2	1	2	3	0	0	3
o_9	3	1	3	1	2	0	2	0
o_{10}	3	1	2	1	2	0	2	0

Generated rules:

1. $C_1(3) \rightarrow D_{12}$,
2. $C_1(1)$ AND $C_3(1) \rightarrow D_{11}$,
3. $C_1(2)$ AND $C_3(3) \rightarrow D_{11}$,
4. $C_1(2)$ AND $C_3(2) \rightarrow D_{13}$,
5. $C_1(1)$ AND $C_4(2) \rightarrow D_{12}$,
6. $C_1(1)$ AND $C_5(2) \rightarrow D_{13}$,
7. $C_3(3)$ AND $C_5(1) \rightarrow D_{11}$.

Computation time: 34.4 ms.

The algorithms implemented in *MATLAB* were carefully tested on problem instances of various sizes. The conducted tests validated the presented methodology.

5.4 Implementation of the Rough Inference System

The proposed system was implemented and tested in MATLAB v.2019b programming environment. Its GUI was created with *App Designer*. The application *Rough reasoning* enables:

- reading, generation, and edition of input decision tables DT of the given size with any sets of attributes and decisions values
- checking if DT is well determined (if not, DT may be adjusted)
- finding minimal information reduct in DT
- generation of the set of rules for a well determined DT
- decomposition of multivalued decision vectors DV into two valued vectors DV
- generation of sets of exact and approximate nonredundant rules for DT

The application *Rough reasoning* has modular structure which enables adding new features. The GUI of the program contains the following tabs: 1. *Generation of the decision table*; 2. *Instantiation of the decision table* (if the table is not well determined); 3. *Attribute reduction*; 4. *Rule generation*; 5. *Generation of exact and approximate rules*; 6. *About the program*. Let us describe two of them.

The *Attribute reduction* tab of the program window is partitioned into three sections: section A contains the loaded input attribute table, section B displays the computed table reducts, while the section C displays the final reduced attribute table.

The *Rule generation* tab of the program window is partitioned into two sections: section A displays the decision table DT while the section B shows the rule set generated by algorithm DT-RULE-GEN (Figs. 5.1, 5.2 and 5.3).

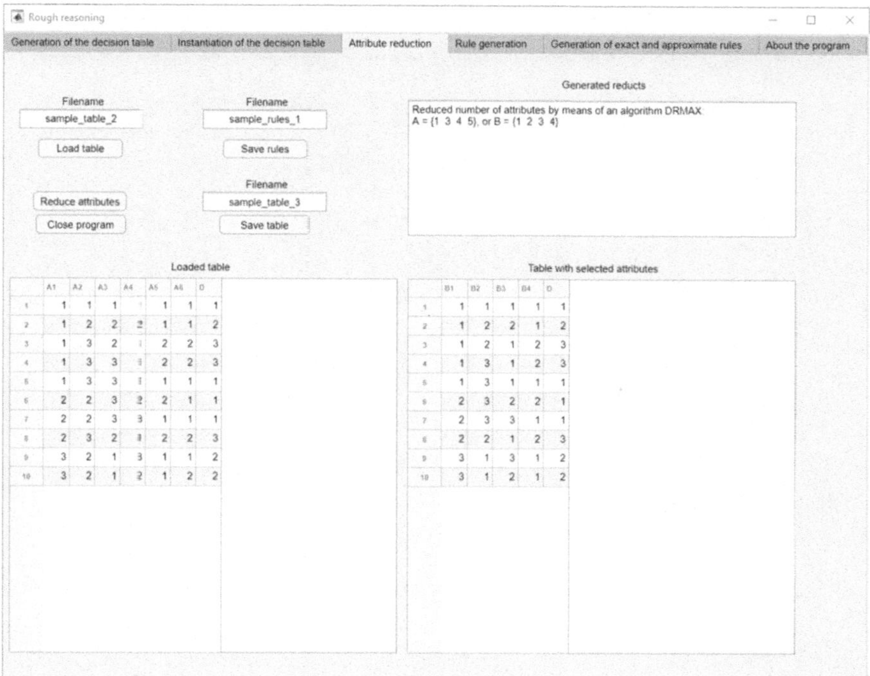

Fig. 5.1 Attribute reduction TAB for Example 3

5.5 An Application in Electrical Engineering—A Case Study

In computer aided diagnosis both admissible and inadmissible states of a system at hand are taken into account. They define into two categories: admissible states correspond to normal functioning of the system while extraordinary inadmissible states reveal abnormal system behaviour. In system monitoring admissible states do not require any action from the supervisor while inadmissible states may generate warnings or alarm calls (interrupts) that must be served. An application in electrical engineering is presented as an illustrative example.

Let our rough inference system be used in computer diagnostics of a transformer state by chromatographic analysis of free and dissolved gases in transformer oil (DGA) [8, 16]. There are two classes of faults: electrical faults and thermal faults. In a simplified example given below the rough inference system uses a set of the input attributes (physical values to be measured) and generates the decision rules for transformer diagnosis.

Example 6 IEC-60599 standard determines conditions for transformer state diagnosis (Table 5.8). The task is to find a simplified diagnostics table for IEC-60599

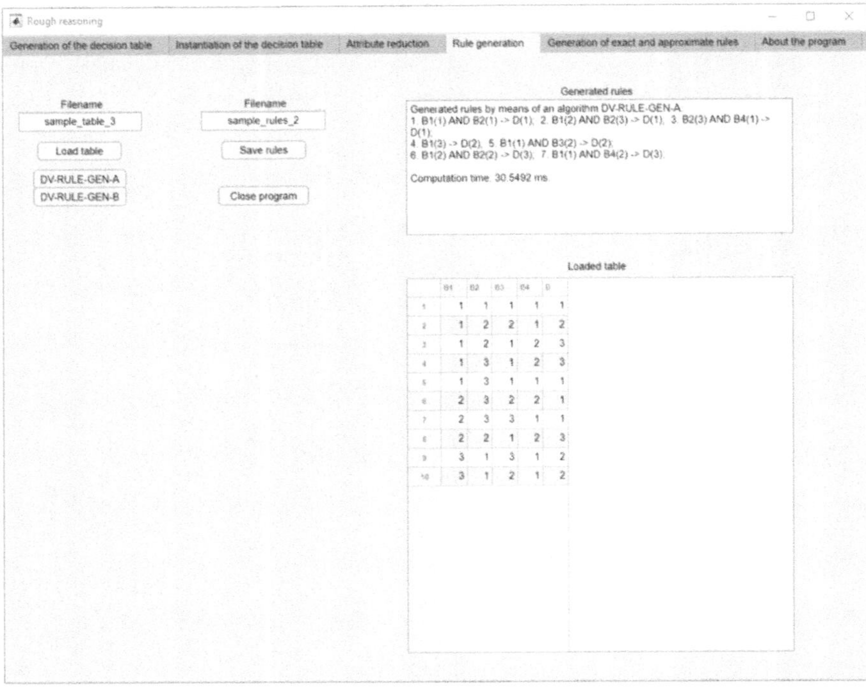

Fig. 5.2 Rule generation TAB for Example 4

interpretation on the basis of rough set based reasoning (cf. [6]).

Selected notations [8]:
C_2H_2—high energy discharge,
CH_4—low temperature fault,
C_2H_4—high temperature fault,
$R1$, $R2$, $R5$—are 3 out of 10 defined gas ratios:
$R1 = CH_4/H_2$; $R2 = C_2H_2/C_2H_4$; $R5 = C_2H_4/C_2H_6$.
PD (partial discharge), $D1$ (low energy discharge), $D2$ (high energy discharge) - are electrical faults,
$T1$, $T2$, $T3$—are thermal faults.

The normalized form of $DT(C, D)$ table is depicted in Table 5.9.

Due to attribute dependencies in C the following indiscernibility classes can be found: for $B = \{C_1\}$ – $\{1, 4\}, \{2, 3\}, \{5, 6\}$; for $B = \{C_2\}$ – $\{1\}, \{2\}, \{3\}, \{4, 5, 6\}$; for $B = \{C_3\} - \{1, 4\}, \{2, 3, 5\}, \{2, 3, 6\}$. Only $B = \{C_2, C_3\}$ has no indiscernibility classes among all three two-column B tables.

Reduction of DT is not performed. The indiscernibility classes are taken into account during execution of DV-RULEGEN-A and DV-RULEGEN-B.

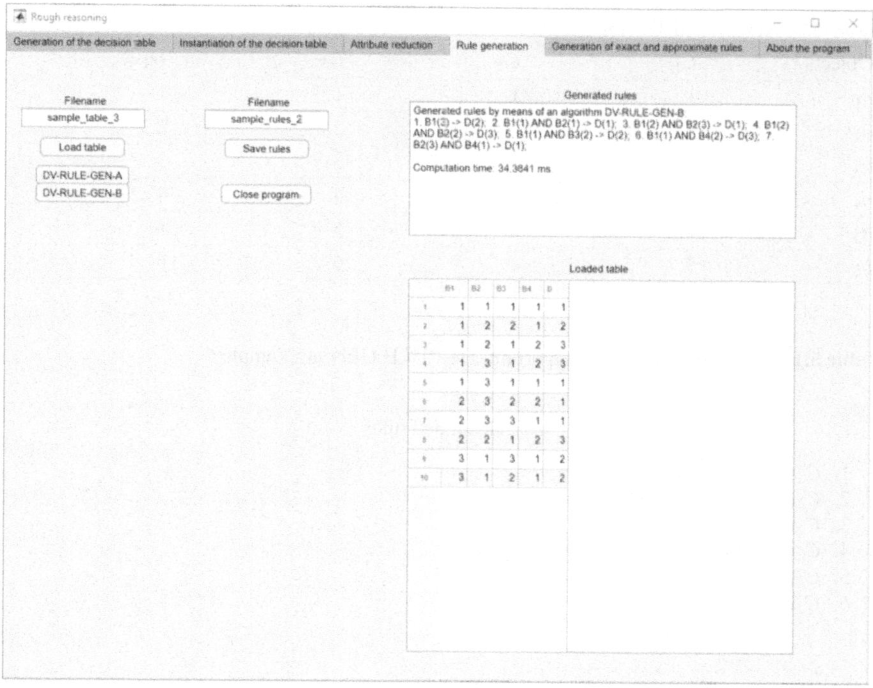

Fig. 5.3 Rule generation TAB for Example 5

Table 5.8 *DT* for Example 6 (IEC-60599)

$R2$	$R1$	$R5$	Diagnosis
NS	< 0.1	< 0.2	Partial discharge (PD)
> 1	0.1–0.5	> 1	Low energy discharge (D1)
0.6–2.5	0.5–1.0	> 2	High energy discharge (D2)
NS	> 1	< 1	T<300 °C (T1)
< 0.1	> 1	1–4	300 °C<T<700 °C (T2)
< 0.2	> 1	> 4	T>700 °C (T3)

In Table 5.10 the set of rules is shown generated by variants of the algorithm DT-RULEGEN. In this particular the algorithm's both variants produce the same set of rules in the same order of generation. □

It seems from Example 6 that the power transformer diagnosis is an easy task. In fact the problem is much more complex because it involves more general faults that can be caused by windings or transformer cores etc. Experiments with neural network classifiers delivered much more detailed data about combinations of attributes for various kinds of faults in the area of transformer diagnostics [16].

Table 5.9 Normalized DT for Example 5

Objects U	C_1	C_2	C_3	D_1
o_1	1	1	1	1
o_2	2	2	2	2
o_3	3	3	3	3
o_4	1	4	4	4
o_5	4	4	5	5
o_6	5	4	6	6

Table 5.10 Rules generated with algorithm DT-RULE-GEN in Example 6

Generated rules

1. $C_2(1) \rightarrow D_{11} = PD$,
2. $C_2(2) \rightarrow D_{12} = D1$,
3. $C_2(3) \rightarrow D_{13} = D2$,
4. $C_2(4)$ AND $C_3(4) \rightarrow D_{14} = T1$,
5. $C_2(4)$ AND $C_3(5) \rightarrow D_{15} = T2$,
6. $C_2(4)$ AND $C_3(6) \rightarrow D_{16} = T3$.

5.6 Conclusions

In this chapter a rough inference software system for computer-assisted reasoning was described. It is useful for processing information systems which can be modeled by means of rough sets.

An original contribution of our research is developing algorithms and a new software tool for fast processing of rough data. The methodology is based on rough set theory. Two original approximate algorithms for processing multivalued decision tables were presented. Several examples were provided for clarification of algorithmic details.

The application example given in Sect. 5.5 provides an insight into modeling and designing a diagnostic system for a real electrical device. During long exploitation of power transformers a number of electrical and thermal faults can occur that should be detected, classified and properly handled. Depending on the required accuracy of the diagnostic system several general AI methods can be applied including neural networks and rough sets.

We believe that the presented results concerning the designed rough inference system justify further development the software by adding next modules and new algorithms.

Acknowledgements This work was partially supported by a research grant from Cracow University Technology.

References

1. R. Bello, Y. Gómez, Y. Caballero, A. Nowe, R. Falcón, Rough sets and evolutionary computation to solve the feature selection problem, in *Rough Set Theory: A True Landmark in Data Analysis. Studies in Computational Intelligence*, vol. 174, eds. by A. Abraham, R. Falcón, R. Bello (Springer, Berlin, Heidelberg, 2009). https://doi.org/10.1007/978-3-540-89921-1_9
2. G. Borowik, T. Łuba, Fast algorithm of attribute reduction based on the complementation of boolean function, in *Advanced Methods and Applications in Computational Intelligence. Topics in Intelligent Engineering and Informatics*, vol 6, eds. by R. Klempous, J. Nikodem, W. Jacak, Z. Chaczko (Springer, Heidelberg, 2014), pp. 25–45. https://doi.org/10.1007/978-3-319-01436-4_2
3. G. Chandrashekar, F. Sahin, A survey on feature selection methods. Comput. Electr. Eng. **40**, 16–28 (2014). https://doi.org/10.1016/j.compeleceng.2013.11.024
4. F.J. Gimpel, A method of producing a Boolean function having arbitrary prescribed prime implicant table. IEEE Trans. Comput. 485–488, C-14 (1965). https://doi.org/10.1109/PGEC.1965.264175
5. M.R. Garey, D.S. Johnson, *Comput. Intractability* (A guide to the theory of NP-completeness, Freeman, San Francisco, 1979)
6. A. Junid, S. Li, L. Ni, Dissolved gas analysis and its interpretation techniques for power transformers, Technical Report, UNSW Sydney (2008)
7. D. Jain, V. Singh, An efficient hybrid Feature Selection model for dimensionality reduction. Int. Conf. Comput. Intell. Data Sci.(ICCIDS 2018), Procedia Comput. Sci. **132**, 333–341 (2018). https://doi.org/10.1016/j.procs.2018.05.188
8. J.R. Jung, H.D. Seo, S.J. Kim, S.W. Kim, Advanced Dissolved Gas Analysis (DGA) diagnostic methods with estimation of fault location for power transformer based on field database, in *Proceedings of CIGRE* (Paris, France, 2016)
9. A. Kapralski, Theory of Boolean spaces and its applications in synthesis of switching circuits, Ph.D. dissertation, AGH University of Science and Technology, Kraków (1979) (in Polish)
10. A. Kapralski, W. Skarbek, Problem of searching minimum base in Boolean tables, in *Podstawy sterowania* (1986), pp. 257–265
11. W.D. Kozakow, Minimization of multi-variable logic functions. Avtomatika i telemechanika **23**(9) (1962) (in Russian)
12. U.M. Khaire, R. Dhanalakshmi, Stability of feature selection algorithm: A review. J. King Saud Univ.–Comput. Inf. Sci. https://doi.org/10.1016/j.jksuci.2019.06.012, in print
13. M. Kryszkiewicz, Rough set approach to incomplete information systems. Inf. Sci. **112**, 39–49 (1998). https://doi.org/10.1016/S0020-0255(98)10019-1
14. G. Lang, M. Cai, H. Fujita, Q. Xiao, Related families-based attribute reduction of dynamic covering decision information systems. Knowl. Based Syst. **162**, 161–173 (2018). https://doi.org/10.1016/j.knosys.2018.05.019
15. H. Liu, Feature selection, in *Encyclopedia of Machine Learning*, eds. by C. Sammut, G.I. Webb (Springer, Berlin, 2010). https://doi.org/10.1007/978-0-387-30164-8_306
16. V. Miranda, A.R.G. Castro, Improving the IEC table for transformer failure diagnosis with knowledge extraction from neural networks. IEEE Trans. Power Deliv. **20**(4), 2509–2516 (2005). https://doi.org/10.1109/TPWRD.2005.855423
17. J. Miao, L. Niu, A survey on feature selection, information technology and quantitative management (ITQM 2016). Procedia Comput. Sci. **91**, 919–926 (2016). https://doi.org/10.1016/j.procs.2016.07.111
18. Z. Pawlak, Rough Sets. Int. J. Inf. Comput. Sci. **11**, 344–356 (1982)
19. Y. Qian, J. Liang, W. Pedrycz, C. Dang, Positive approximation: An accelerator for attribute reduction in rough set theory. Artif. Intell. **174**, 597–618 (2010). https://doi.org/10.1016/j.artint.2010.04.018
20. Dissolved gas analysis interpretation and intelligent machine learning techniques, in *Transformer Ageing: Monitoring and Estimation Techniques*, eds. by T.K. Saha, P. Purkait (Wiley, Singapore Pte. Ltd., 2017), pp. 211–243. https://doi.org/10.1002/9781119239970.ch4

21. A. Skowron, C. Rauszer, The discernibility matrices and functions in information systems. in *Intelligent Decision Support. Theory and Decision Library (Series D: System Theory, Knowledge Engineering and Problem Solving)*, Vol. 11, ed. by R. Słowiński (Springer, Dordrecht, 1992). https://doi.org/10.1007/978-94-015-7975-9_21
22. S. Salesi, G. Cosma, M. Mavrovouniotis, TAGA: tabu asexual genetic algorithm embedded in a filter/filter feature selection approach for high-dimensional data, to appear in Information Sciences
23. D. Ślęzak, S. Stawicki, The problem of finding the simplest classifier ensemble is NP-Hard—a rough-set-inspired formulation based on decision bireducts, in *Rough Sets. IJCRS 2020. Lecture Notes in Computer Science*, vol. 12179, eds. by R. Bello, D. Miao, R. Falcon, M. Nakata, A. Rosete, D. Ciucci (Springer, Cham, 2020). https://doi.org/10.1007/978-3-030-52705-1_15
24. W. Xu, X. Zhang, J. Zhong, W. Zhang, Attribute reduction in ordered information systems based on evidence theory. Knowl. Inf. Syst. **25**, 169–184 (2010). https://doi.org/10.1007/s10115-009-0248-5
25. Y. Zhao, Y. Yao, F. Luo, Data analysis based on discernibility and indiscernibility. Inf. Sci. **177**(22), 4959–4976 (2007). https://doi.org/10.1016/j.ins.2007.06.031

Part II
Advances in Artificial Intelligence-based Applications and Services

Chapter 6
Context Representation and Reasoning in Robotics-An Overview

Konstantinos Dimitropoulos and Ioannis Hatzilygeroudis

Abstract Information (including knowledge) representation is a crucial factor in artificial intelligence and especially in robotics. Information includes data about the environment, its inhabitants and all the objects that are within it, as well as attributes and relations between those objects, including humans. This information constitutes what is called "context" of a robot or a machine and can be represented in various ways. This contextual information can be used to answer queries, and even to create other pieces of relevant information, using reasoning. In this paper, we review basic research work focusing on representation of this kind of information and the reasoning mechanisms that use it to perform inferences and produce (new) results. We focus on the main aspects that concern this research field, that is context definition, representation and reasoning. Finally, we propose a few research directions that could be explored.

Keywords Context definition · Contextual information · Context representation · Context reasoning

6.1 Introduction

Human–robot communication is one of the major research areas of artificial intelligence, with growing importance and interest. The major objective of this research direction is to achieve a fast and effective communication between human and robot. To achieve this, a clear and analytical description of all data that is used for information exchange between human and machine should be available in a proper format. In human–robot interaction, this data may lead to modelling humans and human cognition. A robot must often perform various tasks based on abstract instructions,

K. Dimitropoulos · I. Hatzilygeroudis (✉)
Department of Computer Engineering &Informatics, University of Patras, Patras, Greece
e-mail: ihatz@ceid.upatras.gr

K. Dimitropoulos
e-mail: kdimitrop@upatras.gr

M. Virvou et al. (eds.), *Advances in Artificial Intelligence-based Technologies*, Learning and Analytics in Intelligent Systems 22,
https://doi.org/10.1007/978-3-030-80571-5_6

which the robot interprets in a symbolic form. Usually, there are differences between those instructions and the real (or grounded) information that the robot needs in order to execute the tasks. This gap between the abstract (or symbolic) information and the specific one, which is required for real robot actions, should be bridged via certain knowledge processing methods, appropriate and structured representation of the pieces of knowledge and finally reasoning with them [1]. There are also cases where previous collected data should be studied for the new data to be processed efficiently. So, it seems that the biggest challenges are, how the description of that data will be made and how the communication process will become as reliable as possible. Therefore, the primary goal is the appropriate representation of different types of data on the machine. That data is related to what is called "context" of the robot. Context representation is an area of great importance, the study of which is required for the correct representation of information. Reasoning is related to the mechanics and methods that use that kind of information and produce very interesting results. In this work, we use the term 'information' as including also the term 'knowledge'.

The proper formulation of context representation is very important for a lot of reasons. First, the already known knowledge for a certain entity is enriched with extra useful information, which is something that can lead to better disambiguation. Second, it can offer the unification of different collections of personal beliefs, that may exist at the same time or not. In this way, pieces of knowledge that belong to different perspectives can unify in order to make the set of known information complete. Keeping the above in mind, someone can understand that the human–robot communication becomes easier and much more efficient. This is so because, even if the information that is exchanged is not completely understandable, it is accompanied by an organized and complete mass of data that can provide complete disambiguation and understanding of this "unknown" concept/information. Finally, by using this more complete and advanced representation of information, artificial intelligence systems that use knowledge representation and reasoning mechanisms, become even more efficient.

Context representation and reasoning is a subject with rich research as it is considered a pillar in artificial intelligence. There have also been other surveys trying to register this wealth of research. There is a survey that refers to context representation and reasoning mostly on robotic path planning scenarios [2]. In [3], the authors try to review the research work focusing on acquisition, management and deployment of contextual information. There is also an attempt of registering and comparison of all the available semantic web reasoners and tools [4]. However, there is no survey that combines the work that has been done generally on the concept of context and knowledge representation, as well as the various reasoning techniques and tools that are used in this research field. This is the objective of this overview. To provide a presentation of the concept of context, and its main representation and reasoning methods.

The organization of the paper is as follows: Sect. 2 presents the concept of Context and Knowledge Representation with references to other papers and studies. Section 3 describes the techniques and methodologies that are used in the Reasoning part.

Section 4 is the final section and introduces possible future avenues for research and is followed by references.

6.2 Context

6.2.1 Definitions of Context

In computer science, according to the Online Dictionary of Computing[1]: "context is that which surrounds and gives meaning to something else". More specifically, context may include parts of a conversation, the human perspective or those situations that bring light to the concept of a subject and are connected closely to that. However, there is no official definition of the term context and usually the researchers try to describe one that fits in their needs and implementations.

Definitions of context used in the literature fall into two main categories: The first one describes context based on specific entities (locations or objects), while the second one is based on the relationships and the structure of the surrounding information. A machine that uses contextual information, in order to become more efficient and "smarter", is called context-aware machine. The term context-aware was first used by Schilit and Theimer [5]. At the beginning, location was the primary content of context, but later the notion of the identity of nearby people and objects were also included. Turner [6] defines context as "an identifiable configuration of features which are meaningful to characterize the world state and useful to influence the decision process of a robotic system". He also divided contextual information in three parts, depending on the kind of information it describes: Environmental information. Task-related information and Agent self-knowledge. One of the most complete and generalized definitions states that "Context is any information that can be used to characterize the situation of an entity. An entity is a person, place, or object that is considered relevant to the interaction between a user and an application, including the user and applications themselves" [7, 8]. Later, Kofon and Mikalsen [9] extended that definition to "Context is the set of suitable environmental states and settings concerning a user, which are relevant for a situation sensitive application in the process of adapting the services and information offered to the user" giving an application-oriented flavor.

Referring to specific problems, context is described differently from implementation to implementation. Let's consider a motion planning problem using a robotic ground vehicle, where the robot must move in an environment with obstacles and arrive at its target destination without colliding with the obstacles. In that case, the certain motions that the robot can make, the type of objects in the environment along with their parts, the different states of the robot during the plan execution and environmental regions that enclose the objects and the final destination, are considered

---

[1] https://www.oxfordreference.com/view/10.1093/oi/authority.20110803095834413.

as context [10]. On the other hand, when there is a human in the equation who is present and plays an important role during the application, the contextual information is called "user context". User context is divided into subcategories: environment context, personal context, task context, social context and spatiotemporal context. Environment context captures the entities that surround the user, while personal context provides information about his physiological and mental state. Task context describes the user's goals, tasks, actions and the social context provide information about his social aspects such as friends, enemies, co-workers. Finally, spatiotemporal context is concerned with attributes like time, location and movement [9, 11]. This kind of context is used primarily by systems that are running in a mobile environment.

6.2.2 Context Aware Systems

Context aware systems can adapt to changes in the environment by being aware of the context, which they are running in. Some factors that influence this adaptive behavior are, the location of use, the collection of nearby people, hosts and accessible devices, as well as changes to such entities over time. The benefit that such a system has is that it can monitor and react to changes to the environment [12]. Three important aspects of contextual information that a system should examine in order to be able to recognize those changes are, where you are, whom you are with and what resources are nearby you. So, an updated context aware system needs to make use of the contextual information and exploit the change of context [11].

Schilit and Theimer [5] developed such a system, where users run software that is constantly monitoring or subscribing to information about the world around them. The monitoring is done in order to highlight and to keep a record of interesting located objects, where a person had chosen to make his present location publicly available, as well as persons once had encountered. Also, to detect location-specific information and nearby devices. And, finally, to detect nearby people, located objects or services that are relevant to reminders or actions set to be triggered by their presence.

Another good example of how a context aware system takes advantage of the contextual information is the implementation of an autonomous vehicle [13]. In the case of a non-context aware autonomous vehicle, the path to its destination is provided by static maps and GPS signal and there is no further information provided for the traffic or the weather. However, in the case of the context aware autonomous vehicle, the vehicle has access to data representing the traffic over several roads during different periods of the day, as well as weather condition using an Internet connection. It can also take commands from users regarding the driving mode or the preferred roads, because a button console for acquiring contextual information straight form the users is also available.

6.2.3 *Context Representation*

It is not adequate to simply recognize what could be considered context and what could not and acquire it. The second step is to find ways to represent it effectively. The collected data should be organized and represented in a uniform and appropriate form so that the reasoning mechanisms could process them and generally, for the context aware systems, to take advantage of them. There are three different models for context representation: Embedded, Logic-based and Probabilistic [3].

6.2.3.1 Embedded Representations

In Embedded Representation, context is represented as sets of meaningful sensory features that characterize particular situations. Usually, this kind of representation is used by reactive systems that need to react and recognize the current context and adapt its behaviors quickly, as for example systems performing visual recognition. However, this type of representation is strictly dependent on the perceptual features and as someone can imagine is not the most versatile method for context representation.

Narayanan et al. [14] proposed a robot learning framework to classify and acquire scenario specific autonomous behaviors through demonstration using this kind of representation. The goal was to distinguish among different traversing scenarios and identify the best matching context within the scenario to predict a robot behavior. They proposed a two-level matching scheme. In the first one, a classification of the traversing scenario is performed in order to recognize the type of geometry and shape of the robot's environment. The second one is the more interesting one because it concerns the identification of similar context within that recognized scenario. In this case, context is referred to as a history or a trace of visual perceptions of other recent scenarios. For the two types of context to match, there should be a similarity between the current and the demonstrated perceptual traces and motion trajectories, which characterize the demonstrated one.

Enhanced image assessment is another application of embedded representation. More specifically, the use of multisensor image fusion for surveillance applications. According to this method, a fused image can be generated from multiple sensor inputs. Liu et al. [15] conducted a study on selected image fusion metrics and multiresolution image fusion algorithms and tackled the problem of a night vision application. In order to assess a fused image, two methods were considered. In the first one the fusion result is compared with a known reference image while the second one is a nonreferenced assessment. The latter is the one that the authors of that paper used, along with 12 fusion metrics that had been implemented in Matlab and applied to six multiresolution fusion algorithms.

6.2.3.2 Logic-Based Representations

One of the most famous ways to represent knowledge and more specifically context is the use of Ontology. Ontology based knowledge representation describes the individual instances and roles in the domain that are represented using unary and binary predicates [1]. Usually an ontology is composed of five components: classes (concepts), individuals (instances), relations, functions and axioms. There are editors, like Protégé [16], where ontologies can be created and there are also several ontology markup languages namely RDF-RDFS, OIL, DAMN+OIL, OWL, OWL-DL, OWL-Lite [2]. Logic-based representations range from rule-based ontologies to first order logic.

According to Muhayyuddin et al. [10], the knowledge around task and workspace is organized into two representation levels: High Level knowledge, which is static (abstract knowledge), and Low Level knowledge, which is dynamic (instantiated knowledge). They use ontology models in order to organize those levels of knowledge into specific sectors and to determine objects and the relations between them. At the end they develop a framework and a planning algorithm and solve a motion planning problem. Lemaignan et al. [17] created a knowledge model, which consists of five submodules, and when those were combined they carried out a specific task. In the core of it was a knowledge base, where all the relevant information was stored and queried back when needed. The available knowledge was derived from a priori knowledge, perception, interaction, planning and inferences in symbolic form from the reasoning mechanism, while the knowledge representation was carried out with RDF triplets in OWL language. Furthermore, in order to represent the knowledge in the knowledge base they created their own type of ontology, the OpenRobots, by using the OpenCyk [18] upper ontology. Galindo and Saffiotti [19] tried to infer robot goals from violations of semantic knowledge. They worked on the case where the semantic knowledge turns out to conflict with the robot's observations. In order to describe the objects in the environment and their relationships, they used a semantic map which comprises two different but interconnected parts: a spatial box (S-Box) and a terminological box (T-box). OWL-DL was used as a formalism to represent semantic knowledge, while normative knowledge was represented with symbolic form in description logic.

A different approach to representation, based on first order logic this time instead of ontologies, is the use of Inductive Logic Programming (ILP) [20]. The problem that was tackled here was how robots can use gaining experiences from previous execution failures in order to become more efficient and flexible. ILP in this situation helps on finding, updating or abandoning hypotheses, which need to be consistent with the observed data and was expressed by a set of first order logic sentences. Each hypothesis was associated with a certain weight based on its confidence on correctly modeling observations. Formulas implemented in first order logic were used, in order to describe those hypotheses and reject rules that represented the failed operators.

6.2.3.3 Probabilistic Representations

Because the risk of error and uncertainty exists in any robotic system, a probabilistic representation of the system is often needed, in which relations between context and desired behaviors are formalized through probabilistic structures. The whole representation is based on the identification of contextual variables and the ability of probabilistic structures, like Bayesian Networks, to model the degree of belief of the different scenarios, as well as the most likely behavior that the system will have in order to adapt to the new identified context.

This kind of representation could be very beneficial to the field of share autonomy, in which a function that is required by a robot is partially or fully executed by a human. In [21], such an approach is described, in which a user is able to specify contextual information related to the specific task of the appropriate grasp of an object by a robot, in order to help the robot execute. The main procedure is the following: the robot autonomously generates a set of viable grasps and then uses the user provided constraints to select a certain grasp. In this approach, Witzig et al. define context based on a set of features, which are either generated automatically or provided by the user through a novel dialog-based user interface. A Bayesian Network [22] with discrete and continuous variables is used in order to encode the contextual information into a probability density function. First, the context provided by the user is combined with a context derived from online perception. After that, the probability density function is calculated to sort the set of candidate grasps. Finally, the most effective grasp pose is selected.

Another effort, this time with the help of context and scale, tries to solve the problem of car detection in images [23]. In that paper, a probabilistic model for combining appearance, scale and context scores for object detection in images is presented. The appearance model that was used was Felzenszwalb's deformable parts-based model [24]. First, this model run on a scale-filtered image and the result was a set of bounding boxes that came up with a score which represented how much the appearance of this bounding box matched with that of a car. Because there was a chance of false detections, two scale scores were computed. The first one estimated the real-world height of an object within a bounding box in the image and the second one the variance caused by errors in the previous height estimation. There were also two context scores that considered the fact that a car might not be always in the middle of the road as well as false estimations about which parts of the image contained the road. The image was filtered based on scale and in the final step all the above were used for a new prediction for each bounding box to be computed.

6.2.3.4 Advantages and Disadvantages

As is commonly recognized, each representation technique is useful and effective on certain applications. There is no perfect type of context representation that can cover any need and all of them have their strengths and weaknesses. Embedded representations try to achieve effective mapping between the sensory data and the estimates for the contextual variables, but the information that they produce is sometimes not fully understandable. Logic representations on the other hand have expressiveness and can effectively describe effects and causes of contextual scenarios, but they do not support powerful enough reasoning tools that can support more complex computations. In addition, some languages that are used to develop ontologies are more descriptive than others. For example, languages like OWL (Web Ontology Language) [25] are limited to the use of binary and unary relationships, while NRKL (Narrative Knowledge Representation Language) [26] uses n-ary ontologies which provide an even better type of expressiveness. Finally, probabilistic representations don't have this type of expressiveness, but they offer reliability on their estimates. Although such efforts improve context representation, more advanced combinations of single representations are required, especially for more complex problems.

6.2.3.5 Context Middleware and Combined Approaches

In the literature, the use of a Context Middleware can solve several problems, where context representation is needed. One of them was how to provide the applications in the mobile environment with relevant contextual information so that they could respond faster and more complete to the user. Kofon and Mikalsen [9] made an approach to automatic situation assessment in a mobile environment within the AmbieSense [27] research project. In their approach they relied on a middleware that provides applications with dynamic access to contextual information, that is relevant with their execution. They also extended the definition of context given by Dey [8] along with the creation of context templates, so that the context middleware could distinguish what information was considered context for a certain application and what was not. Another problem that was tackled was how the different types of contextual information could combine with each other. Bloisi et al. [3] managed to combine the three types of Turner's context knowledge classification [6] on a context-aware autonomous vehicle scenario. They developed a framework for information fusion that was structured into three main phases: Acquisition of the sensor data, Reasoning and Decision where the contextual information was used to adapt the system configuration in accordance with the current context. The Context Middleware that was also used, translated the inferred contextual knowledge in a suitable format for the underlying system. As far as the implementation of the information fusion is concerned, the autonomous vehicle could adapt its speed according to the road conditions (environmental information), the driver's needs (task-related information) and the vehicle condition (agent self-knowledge).

KnowRob [28] is a robot knowledge processing system. It combines information of many types from different sources and several inference methods. It uses its own ontology, the KnowRob ontology, which contains about 7000 object classes and offers powerful reasoning tools that can produce inferences based on this ontology structure. It also supports other reasoning algorithms that use any kind of computations but with one restriction, they must read their input from the knowledge base and produce facts formulated in the common representation language. Usually, this language is OWL. The creators of KnowRob, in another paper [1], did exactly that, they used OWL ontologies as the representation language and combined the inference mechanisms that are integrated in OWL with those that the Knowrob offers. They used SWI Prolog [29] as a programming language and its Semantic Web Library as a tool to load the OWL ontologies into the system. The structure of the ontologies was represented as Prolog predicates. Muhayyuddin et al. [10] also took advantage of the abilities of this system in their motion planning implementation. Two different reasoning methods were used there, reasoning about the manipulation actions and reasoning about the geometry of the objects. They also tried a similar approach by using reasoning predicates in Prolog and then they integrated them within Knowrob.

Context representation can also be applied to cognitive assistance of dependent people at home and more specifically using cloud-assisted robots for the improvement of decision taking. Ayari et al. [30] implemented a cognitive architecture that allowed sharing of storage and communication resources between entities that were connected to the cloud. The ontology that represented the commonsense knowledge was on the main server that was shared between all the entities. They represented spatiotemporal knowledge and the dynamic environment using semantic graphs and NRKL language while the semantic and geometric information of each entity was represented separately. Finally, they introduced a model that extended the representation abilities of the NKRL language for spatial relationship representation and for entities properties representation.

6.3 Context Reasoning

6.3.1 Reasoning Approaches and Techniques

Reasoning mechanisms use knowledge in order to infer new conclusions. The knowledge reasoning techniques that are used the most in robotics can be divided into three categories, each one of which includes several techniques [2].

The first one is Ontology Reasoning, in which ontology performs reasoning to infer logical conclusions using its set of asserted facts and axioms. It includes Relation-based Reasoning techniques, which use the relations in the ontologies to infer new conclusions.

The second one is Temporal Reasoning, which is based on time interval and point logics. Temporal reasoning specifies the rules and symbolism for representing and reasoning using propagation matrix [31] for temporal relations. This kind of reasoning is used frequently on robot path planning problems.

Finally, there is Spatial Reasoning [32], which generates the generic knowledge about the entity and spatial relations between entities in the environment. This kind of reasoning can be done by reasoning on hierarchy, spatial relations and fuzzy representation. The reasoning on hierarchy, is based on the hierarchical representation. According to this method, there are nodes and leaf nodes where each parent node represents more general concepts with the form of a set of hypotheses, while the leaf nodes represent single hypotheses about specific individual items. In addition, big role in the development of this kind of reasoning plays the Dempster-Shafer theory for uncertain reasoning [33]. A basic concept of this theory is the frame of discernment, which is the set of all possible singleton hypotheses in the domain. A basic function in the theory, the basic belief assignment, can be used to describe how a sensorimotor feature supports one or more sets of hypotheses to certain degree. This theory also uses the Dempster's Rule of Combination in which the rule combines two basic belief assignments into one basic belief assignment.

Another technique that is used in combination with context representation is Case Based Reasoning [34]. Case-Based Reasoning is concerned with adapting to new situations by remembering similar earlier experienced situations (cases). An example of a work that uses Case Based Reasoning is that of Kofon and Mikalsen [9]. They used this technique because they wanted to reuse past relevant information about a case, which goal was the same or similar with the current one. What they did was retrieving a known case accompanied by the contextual information along with the goal that it achieved and tried to classify the current case so that it matches the retrieved one. After that and when the current associated goal was achieved, the case was stored in a form of a triplet consisted of the contextual information, the problem and the solution.

Another interesting approach is the definition of a set of rules that consisted of one or more conditions and had as a result a new spatial relationship [30]. Depending on the kind of the spatial relationship, there were three different reasoning methodologies: topological, orientation and proximity relationships.

Finally, an example of a reasoning mechanism that uses and produces contextual knowledge with the help of a Context Middleware is that of the framework which Bloisi et al. [3] introduced. The Context Reasoner, during the reasoning phase, relied on contextual information that the Context Middleware provided and was responsible for inferring new contextual knowledge and for making it available back to the Context Middleware. After that, the Middleware informed the reasoner about taken decisions from the system and this information was used to update contextual knowledge.

6.3.2 Reasoning Tools

There are various tools used for context representation and reasoning. Among them, ontology engineering tools, like ontology editors and OWL (or DL) reasoners are the most popular in context-aware systems. The following tools and reasoners are important to set up a practical context for the applications and provide indications for further engagement with context reasoning in robotics [35].

Pellet [36] is a third generation complete OWL-DL reasoner with extensive support for reasoning with individuals, user-defined datatypes and debugging support for ontologies. It offers several features including conjunctive query answering, rule support, axiom pinpointing and various interfaces to make it easily accessible to users. Pellet is implemented in Java and is open source. Lemaignan et al. [17] chose to use this specific reasoner to reason on the knowledge base in combination with other reasoning algorithms like Common Ancestors, Different Ancestors and Clarification and Discrimination algorithms. The reasoner Pellet was also used on the "violation of semantic knowledge" [19] problem, by providing through its output the violated norm and the culprit of the violation.

Other third generation reasoning models that are widely used are RACER [37] and FaCT++ [38]. RACER (Renamed ABoxes and Concept Expression Reasoner), also known as RacerPro, supports various optimization techniques that deal with number restrictions and ABoxes. ABoxes are part of Description Logic knowledge bases and contain knowledge about individuals and the roles that connect them. FaCT (Fast Classification of Terminologies) works as a description logic classifier and supports model logic satisfiability testing. FaCT++ uses the same description logics algorithm as in FaCT but has a different internal structure.

A couple of the most recently developed reasoners are CEL [13] and HermiT [39]. They provide interactive shell interface with complete reasoning supports in polynomial classification service time [4]. CEL (Classifier for EL) is a reasoner for the small description logic EL, which can be used for the computation of the complete subsumption hierarchy between all concepts induced by EL ontologies. Its most distinctive feature is that it can process very large ontologies in reasonable time. HermiT is the first reasoner that is written using OWL. Given an OWL file, HermiT can determine the consistency of the ontology and identify the hierarchical relationships between classes. It is based on a hypertableu calculus and allows lightning fast ontology classification.

When processing OWL Ontologies, the most widely used open source tool is the OWL API [40]. It is a Java API and reference implementation for creating, manipulating and serializing OWL Ontologies. It constitutes the foundation for a range of software tools, including the Protégé ontology editor. A very recent developed library that is based on OWL API is ROBOT [41]. ROBOT is an open source library and command-line tool for automating ontology development tasks. It is written in Java and is compatible with all ontology syntaxes that OWL API supports. One of ROBOT's most important operations is reasoning. The command that is used for reasoning covers two cases: logical validation and automatic classification. ROBOT initiates the reasoner of user's choice using the OWL API Reasoner interface and checks if the ontology is consistent. If it is, then an automatic classification is performed and the direct inferred "subClassOf" axioms are added to the ontology.

Other than OWL (or DL) important semantic web language is SWRL (Semantic Web Rule Language) [42]. It can be used to express rules as well as logic, combining OWL DL or OWL Lite with a subset of the Rule Markup Language [43]. In this language, the set of OWL axioms is extended to include Horn-like rules. Horn-like rules are of the form of an implication between antecedent (body) and consequent (head) and can be read as: whenever the conditions specified in the antecedent hold, then the conditions specified in the consequent must also hold. As far as the reasoning part is concerned, there is an SWRL plugin for Protégé, SWRL-IQ (Semantic Web

Rule Language Inference and Query tool) [27] that allows users to edit, save and submit queries to an underlying inference engine based on XSB Prolog. Among other features, it supports goal-oriented backward-chaining reasoning and more declarative rules and queries. It can trace and debug inference results and then export them in XML or CSV format. SWRL-IQ also has powerful SWRL extensions and a Java procedural attachment mechanism.

6.4 Conclusions and Future Work

The work presented here is related to research in the area of knowledge representation, more specifically in the area of context representation and reasoning. This is an area of artificial intelligence that plays an important role in the human–machine/robot communication and interaction. This work is an overview of the efforts that have been made in the area. The overview deals with what has been done as far as context definition, representation and reasoning are concerned. It is not a complete survey of efforts, but an overview of the main efforts and techniques that are used. Based on them we can also explore possible need for new research directions.

It seems that there have been major steps in the above direction, aiming at an efficient and complete representation of contextual knowledge in the knowledge base of a machine for effective communication and interaction with the human part. However, this is a still growing area, so its objectives are continually extended and there is still room for new steps.

A first aspect is the definition of the notion of "context". It seems that no general commonly accepted definition has been established. Existing definitions are rather application-specific, although there are efforts for generalization. Still, none can be applied to (or cover) all different context situations. This difficulty comes from the fact that contexts in different applications may be quite different.

Frequently, the selection of an information as context includes the acquisition of a set of beliefs that corresponds to a situation and the temporary replacement of those current beliefs with new ones. This procedure may continue every time that new information becomes available and is relevant to a subject. However, at any time and any place more than one context can coexist at different levels, which however all refer to a certain situation. The objective is to carefully craft a context from several atomic ones instead of retrieving only one set of beliefs. There are not a lot of known techniques for representation of such overlapping pieces of knowledge, nor algorithms that could retrieve or discard a piece of information after reasoning on whether it belongs to context or not.

As far reasoning is concerned, more work can be done in the area of context-sensitive applications. There is a method that has recently been used and tested and found to produce good results, which is case based reasoning. But, as this is an ongoing research, the results are still preliminary and there are issues that should be resolved. First, it is a method that needs some kind of verification of the results and suggestions that it produces. Second, how it can be improved further in order to produce even more relevant and influencing solutions that cover the current needs of the user as completely as possible.

References

1. M. Tenorth, M. Beetz, Representations for robot knowledge in the KnowRob framework. Artif. Intell. **247**(2017), 151–169 (2017)
2. R. Gayathri, V. Uma, Ontology based knowledge representation technique, domain modeling languages and planners for robotic path planning: a survey. ICT Express **4**, 69–74 (2018)
3. D.D. Bloisi, D. Nardi, F. Riccio, F. Trapani, Context in robotics and information fusion, in *Context-Enhanced Information Fusion. Advances in Computer Vision and Pattern Recognition*, eds. L. Snidaro, J. García, J. Llinas, E. Blasch (Springer, Cham), https://doi.org/10.1007/978-3-319-28971-7_25
4. A. Khamparia, B. Pandey, Comprehensive analysis of semantic web reasoners and tools: a survey. Educ. Inf. Technol. (2017)
5. B.N. Schilit, M.M. Theimer, Disseminating active map information to mobile hosts. IEEE Network **8**(5), 22–32 (1994)
6. R.M. Turner, Context-mediated behavior for intelligent agents. Int. J. Hum Comput Stud. **48**(3), 307–330 (1998)
7. A.K. Dey, Providing architectural support for building context-aware applications. Ph.D. Thesis, Georgia Institute of Technology, 2000
8. A.K. Dey, Understanding and using context. Pers. Ubiquit. Comput. **5**, 4–7 (2001)
9. A. Kofon-Petersen, M. Mikalsen, Context: representation and reasoning -- representing and reasoning about context in a mobile environment. d'Intelligence Artif. **19**(3), 479–498 (2005)
10. G. Muhayyuddin, A. Aliakbar, R. Jan, Ontological physics-based motion planning for manipulation, in *IEEE 20th Conference on Emerging Technologies Factory Automation (ETFA)* (2015)
11. H.I. Myrhaug, Towards life-long and personal context spaces, in *8th International Conference on User Modeling, Workshop on Context-Aware Applications* (Sonthofen, Germany, 2001)
12. B.N. Schilit, N. Adams, R. Want, Context-aware computing applications, in *1994 First Workshop on Mobile Computing Systems and Applications* (Santa Cruz, California, USA, 1994), pp. 85–90
13. F. Baader, C. Lutz, B. Suntisrivaraporn, CEL – a polynomial – time reasoner for life science ontologies, in *Conference: Proceedings of the Third International Joint Conference on Automated Reasoning* (2006)
14. K.K. Narayanan, L.F. Posada, F. Hoffmann, T. Bertram, Scenario and context specific visual robot behavior learning, in *IEEE International Conference on Robotics and Automation Shanghai International Conference Center* (Shanghai, China, 2011)
15. Z. Liu, Z. Xue, E. Blasch, R. Laganiere, Objective assessment of multiresolution image fusion algorithms for context enhancement in night vision: a comparative study. IEEE Trans. Pattern Anal. Mach. Intell. **34**(1), 94–109 (2012)
16. https://protege.stanford.edu/
17. S. Lemaignan, M. Warnier, E.A. Sisbot, A. Clodic, R. Alami, Artificial cognition for social human–robot interaction: an implementation. Artif. Intell. **217**, 45–69 (2016)
18. D.B. Lenat, R.V. Guha, K. Pittman, D. Pratt, M. Shepherd, Cyc: toward programs with common sense. Commun. ACM **33**(8), 30–49 (1990)
19. C. Galindo, A. Saffiotti, Inferring robot goals from violations of semantic knowledge. Robot. Auton. Syst. **61**, 1131–1143 (2013)
20. S. Karapinar, S. Sariel-Talay, P. Yildiz, M. Ersen, Learning guided planning for robust task execution in cognitive robotics, in *AAAI Workshop on Intelligent Robotic Systems* (2013)

21. T. Witzig, M.J. Zöllner, D. Pangeric, S. Osentoski, R. Jäkel, D. Rüdiger, Context aware shared autonomy for robotic manipulation tasks" in *IEEE/RSJ International Conference on Intelligent Robots and Systems* (Tokyo, 2013), pp. 5686–569
22. J. Pearl, *Probabilistic Reasoning in Intelligent Systems: Networks of Plausible Inference*, Morgan Kaufmann (1988)
23. D. Held, J. Levinson, S. Thrun, A probabilistic framework for car detection in images using context and scale", in *IEEE International Conference on Robotics and Automation River Centre* (Saint Paul, Minnesota, USA, 2012)
24. P.F. Felzenszwalb, R.B. Girshick, D. McAllester, D. Ramanan, Object detection with discriminatively trained part-based models. IEEE Trans. Pattern Anal. Mach. Intell. **32**(9), 1627–1645 (2010)
25. https://www.w3.org/TR/owl2-syntax/
26. G.P. Zarri, *Representation and Management of Narrative Information – Theoretical Principles and Implementation* (Advanced Information and Knowledge Processing. Springer, Berlin, 2009)
27. https://ambiesense.com/
28. M. Tenorth, M. Beetz, Knowrob knowledge processing for autonomous personal robots, in *IEEE/RSJ International Conference on Intelligent Robots and Systems* (St. Louis, MO, 2009), pp. 4261–426
29. https://www.swi-prolog.org/
30. N. Ayari, A. Chibani, Y. Amirat, G. Fried, Contextual knowledge representation and reasoning models for autonomous robots, in *AAAI 2017 Fall Symposium Series* (2017)
31. J.F. Allen, Planning as temporal reasoning, in *Proceedings of the Second International Conference on Principles of Knowledge Representation and Reasoning* (Cambridge, MA, 1991)
32. C. Zetzsche, J. Wolter, K. Schill, Sensorimotor representation and knowledge-based reasoning for spatial exploration and localization. Cogn. Process. **9**(2008), 283–297 (2008)
33. G. Shafer, *A Mathematical Theory of Evidence* (Princeton University Press, 1976)
34. A. Aamodt, E. Plaza, Case-based reasoning: foundational issues, methodological variations, and system approaches. AI Commun. **7**, 39–59 (1994)
35. P. Maier, Modelling and reasoning about robot knowledge with OWL Ontologies, Thesis, 2017
36. E. Sirin, B. Parsia, B.C. Grau, A. Kalyanpur, Y. Katz, Pellet: a practical OWL-DL reasoner. J. Web Seman. **5**(2007), 51–53 (2007)
37. K. Bhoopalam, V. Haarslev, *Fire – Extending Racer by an Engine for SWRL Rules* (Concordia University, Montreal, Canada, 2005)
38. D. Tsarkov, I. Horrocks, *FaCT++ Description Logic Reasoner: System Description* (The University of Manchester, Manchester, UK, School of Computer Science, 2006)
39. B. Glimm, I. Horrocks, B. Motik, G. Stoilos, *Hermi T: Reasoning with Large Ontologies* (Oxford University, Computing Laboratory, 2009)
40. http://owlcs.github.io/owlapi/
41. R.C. Jackson, J.P. Balhoff, E. Douglass, N.L. Harris, C.J. Mungall, J.A. Overton, ROBOT: a tool for automating ontology workflows. BMC Bioinf. **20** (2019)
42. I. Horrocks, P.F. Patel-Schneider, H. Boley, S. Tabet, B. Grosof, M. Dean, SWRL: a semantic web rule language combining OWL and RuleML, W3C Member Submission, 2004
43. H. Boley, B. Grosof, S. Tabet, RuleML-semantic web rule markup language, W3C Working Note Draft (2005)
44. https://protegewiki.stanford.edu/wiki/SWRL-IQ

Chapter 7
Smart Tourism and Artificial Intelligence: Paving the Way to the Post-COVID-19 Era

Aristea Kontogianni, Efthimios Alepis, and Constantinos Patsakis

Abstract In these last years, we are witnessing the constant evolution of technologies such as artificial intelligence, machine learning, blockchain, IoT, and cloud computing. At the same time, before the COVID-19 crisis, the tourism sector steadily increased its growth yearly, raising to be one of the dominant sectors worldwide, generating a wide number of job openings in a plethora of industries. Gradually, the need to bridge technology with tourism, gave birth to the *Smart Tourism* concept, with an increasing number of researchers focusing on it. Artificial intelligence that has already been exploited in many fields could not leave the smart tourism sector unaffected. Actually, based on the realisation that AI could be a game-changer when it comes to the tourism industry, we aim at presenting how AI can be leveraged to bring smartness into tourism. We focus on approaches presented by both researchers and practitioners in the tourism industry, and we analyse the exploited technologies. Now more than ever, with the COVID-19 pandemic, the tourism industry needs to be transformed, destinations need to become "safe" and tourists need to be able to travel again. That being the case, we also attempt to recognise the already exploited AI approaches in this field, as well as ways of utilising technology to resume travel and reboot tourism worldwide safely.

Keywords Smart tourism · Artificial intelligence · Deep learning · Augmented reality · Autonomous agents

A. Kontogianni · E. Alepis (✉) · C. Patsakis
Department of Informatics, University of Piraeus, Piraeus, Greece
e-mail: talepis@unipi.gr; talepis@gmail.com

A. Kontogianni
e-mail: akontogianni@unipi.gr

C. Patsakis
e-mail: kpatsak@unipi.gr

M. Virvou et al. (eds.), *Advances in Artificial Intelligence-based Technologies*, Learning and Analytics in Intelligent Systems 22,
https://doi.org/10.1007/978-3-030-80571-5_7

7.1 Introduction

In the last decade, we have witnessed a promising advancement of information and communication technologies [65] with new technologies such as AI, machine learning, IoT, blockchain, augmented reality and cloud computing assisting to significantly improve life quality [39] and being exploited in a plethora of sectors. Simultaneously, the tourism industry remains one of the largest industries worldwide [10, 42], with many economies relying on it. More specifically, in global scale tourism directly contributed to GDP approximately 2.9 trillion U.S. dollars in 2019 [59], slightly more than 2018. The importance of this sector is further emphasised by the fact that it incorporates a wide number of industries like accommodation, transport, attractions and more [59], generating a plethora of job openings and business opportunities [24, 42].

At the same time, we are living in an era overflowing with tourist-related data regarding accommodation, food, sightseeing etc. That being the case, the need to exploit all available means to understand travellers' preferences and offer personalised services to them is evident. The aforementioned need gradually led to the "birth" of the Smart tourism concept, where digital means are leveraged to enhance the whole tourism experience. Based on [27, 32], *Smart Tourism* can be characterised as the progression from traditional tourism via the exploitation of information technologies. The technological advancements along with the fact that travelling is "made easy" has given birth to this new concept, with more destinations and stakeholders attempting to embed "smartness" in their services [41]. Authors in [42], recognising the need to further build upon Smart Tourism established knowledge, after extensive literature review, have identified and presented eleven key concepts/approaches regarding smart tourism as illustrated in Fig. 7.1.

Artificial intelligence, already successfully exploited in a plethora of fields, could actually be a game-changer when it comes to the smart tourism industry. That being the case, several papers have been published which contribute theoretically or practically to the adoption of AI in smart tourism. However, at the time of writing, the usage of AI in the tourism context does not meet its potential. Now, more than ever, with the COVID-19 pandemic, the tourism industry needs to be transformed in ways that have not been yet explored, to recover from the economic and cultural crisis that precedes this pandemic. Recognising the already exploited AI approaches in this field, as well as ways of utilising technology in the attempt to resume travel and reboot tourism worldwide safely is the primary goal of this research. Thus, after reviewing a vast number of articles that focus on smart tourism and artificial intelligence, we present the more prevailing approaches in the field along with future directions and challenges. At the same time, articles published during the COVID-19 pandemic that hold the promise to contribute to the global effort to confront this virus via technological means are reviewed, and the potential impact of them in the smart tourism context is also discussed.

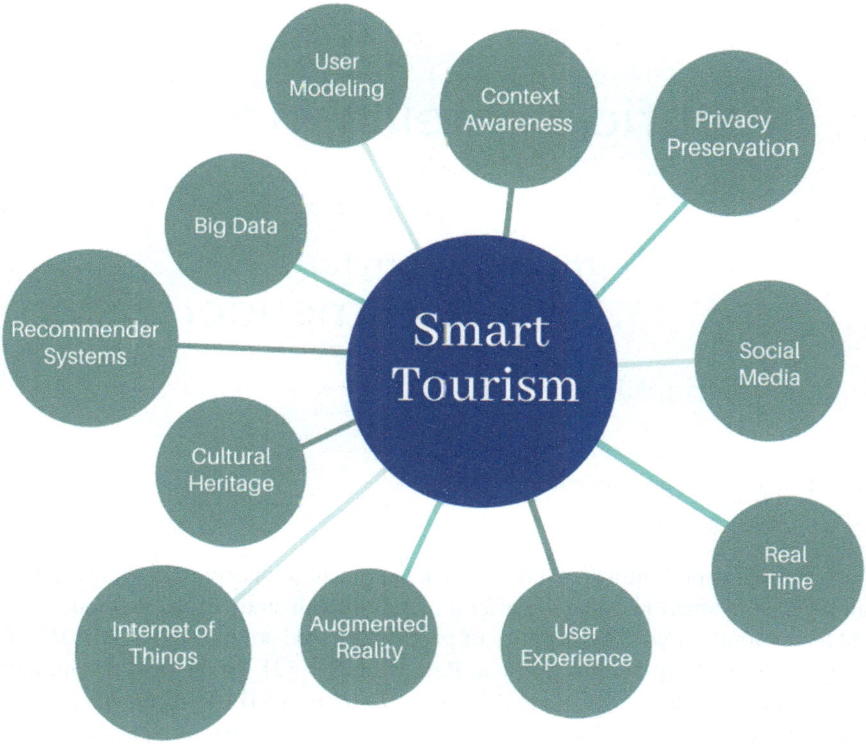

Fig. 7.1 Smart tourism concepts/approaches

7.2 Methodology and Research Approach

One of the main goals of this research is to present artificial intelligence approaches in the context of smart tourism. To achieve that, we have reviewed a vast number of research papers published in the past years while, at the same time, we have located AI tools exploited by key stakeholders in the tourism industry to embed "smartness" in their services. Doing so, we manage to present the road so far regarding artificial intelligence approaches in the smart tourism context and set the ground for further exploitation of artificial intelligence in this field. Three main AI pillars are identified and presented, namely *AI Smart Tourism Recommender Systems*, *Augmented Reality* and *Autonomous Agents*.

After carefully reviewing a wide number of smart tourism related papers, we focused on those who leverage artificial intelligence in the tourism context. More specifically, Fig. 7.2 presents the keywords included in the selected articles. The most frequently identified keywords are highlighted, emphasising the relevance between smart tourism and artificial intelligence in the included articles. To streamline the content, articles which only referred to the possible exploitation of AI in smart tourism or that did not detail the AI techniques/methods exploited were excluded from this

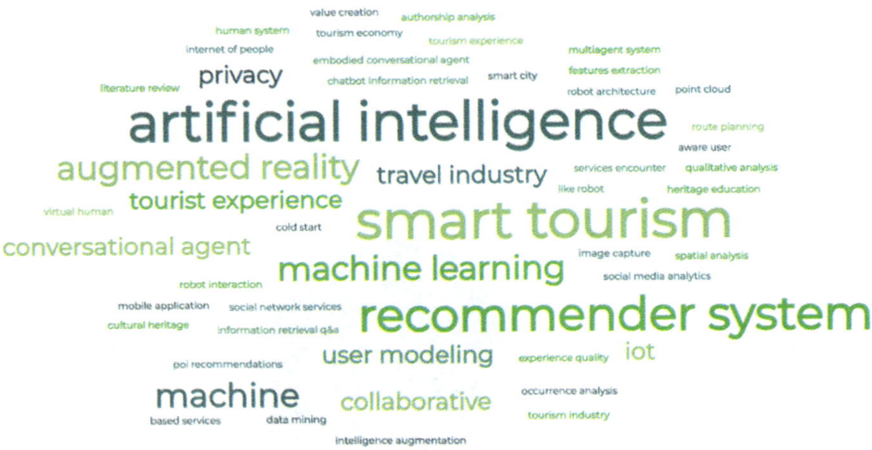

Fig. 7.2 Keywords in reviewed research

work. Further to presenting the most popular approaches/concepts regarding artificial intelligence in smart tourism, we offer a more in-depth study of the methodologies and technologies used. The majority of papers reviewed are indexed in DBLP [20], Google Scholar [55], Scopus [57] and ResearchGate [52]. To this end, we initially queried these databases for the terms: *Smart Tourism* **AND** (*Artificial Intelligence* **OR** *Machine Learning* **OR** *Deep Learning*). We examined the returned titles and the corresponding abstracts. Based on their relevance and quality, we identified the most suitable articles and then used these articles to extract from their references further relevant articles to extend our survey further.

The high quality of computer science journals, proceedings, conference papers, books and thesis in the mentioned above websites in combination with the fact that researches included are conducted by prominent scientists in the field, ensure the reliability and validity of concepts presented. Moreover, inspired by this [31] call for transformative research in the e-Tourism field, we aim at presenting technological solutions that could be exploited in the smart tourism context in the post-COVID-19 era to help tourism survive in this new, totally unexpected reality.

The methodological approach followed is described in the steps below:

1. Identify the need to thoroughly present AI techniques in the smart tourism context as they have been recorded in the international literature
2. Identify the need to locate and document AI technologies utilised worldwide in the tourism industry
3. Recognise the need to document technological approaches that could realise the safe rebooting of tourism in the post-COVID-19 era
4. Select relevant studies, including "grey literature" located through extensive electronic searches, evaluate them and extract identified important data
5. Report the results and the challenges that have arisen from the process mentioned above

7.3 Artificial Intelligence and Smart Tourism

Smart tourism and artificial intelligence are two concepts that combined have the potential to transform the tourism industry completely. This comes as no surprise, as AI paves the way for making autonomous human-like decisions, which in the travel industry facilitates a plethora of actions from personalised recommendations to autonomous agents that respond to tourists. In this section, we aim at presenting how AI is exploited in the smart tourism context by both researchers and practitioners. The aim of AI concepts identified and explained in this work is illustrated in Fig. 7.3.

7.3.1 Artificial Intelligence

The term artificial intelligence, or AI for short, even though has caused a lot of discussion over the years [40], could be defined as the creation of human-like intelligence that attempts to mimic the way a person can learn, perceive, have logic and be capable of performing intellectual processes a human being performs. Artificial intelligence, in other words, refers to the ability of a machine to reproduce human cognitive functions [56] and "make decisions" without human intervention. The artificial intelligence field has a long history as it dates back to 1950 [54]. Since then, AI has been exploited in numerous fields such as medicine, robotics, economy, education, etc. The main subfields of artificial intelligence that have been in the spotlight

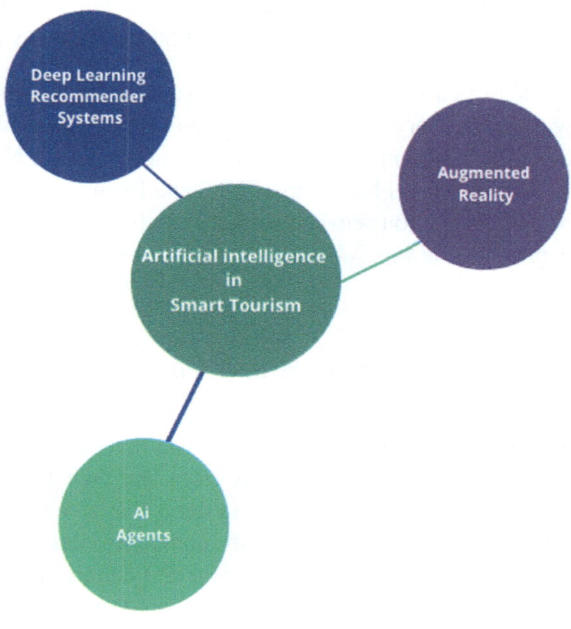

Fig. 7.3 Artificial intelligence and smart tourism

for years now are knowledge representation, natural language processing, expert systems, neural networks, robotics, computer vision and machine learning. More precisely, the main approaches of AI according to [23] are narrow AI, general AI, and super AI.

- **Narrow AI**: Refers to an explicitly defined task that needs to be automated
- **General AI**: This approach aims at the development of a system that mimics the physical and intellectual capabilities of the human mind
- **Super AI**: In contradiction to General AI, the goal of this approach is to create a machine smarter than humans. How something like this is achievable, though, is under discussion

Other commonly used terms that appear in this paper too are *Machine Learning* and *Deep Learning*. The machine learning term, or ML for short, refers to the part of artificial intelligence that provides us with statistical tools for data processing and analysis. ML, contrary to traditional programming, "discovers" rules to perform a data-processing task based on the expected outcomes [18]. Machine learning approaches can be distinguished to the following:

- **Supervised Learning**: The machine learning model exploits labelled data to get trained on mapping an input to an output.
- **Unsupervised Learning**: Data given as input in the model has no "labels" (unlabeled data), and the output is unknown. In such ML models, clustering techniques are commonly used as the model attempts to exploit data "similarity" to group them. K-means algorithm is often used in this machine learning approach.
- **Reinforcement Learning**: In this machine learning approach, the model is trained through a reward system, it must "learn" through trial and error [38].
- Semi-supervised learning Utilises both labelled and unlabeled data during training.

Deep Learning is considered a subfield of Machine Learning which emerged with the primary goal of teaching a machine to "think" like a human being. Main techniques exploited in deep learning are Artificial Neural Networks (ANN), Convolutional Neural Networks (CNN) and Recurrent Neural Networks (RNN).

The correlation between artificial intelligence, machine learning and deep learning is presented in Fig. 7.4.

7.3.2 AI Smart Tourism Recommender Systems

The abundance of tourism-related data that derive from a plethora of sources like explicitly from tourists, their devices and their transactions [4], there is a strong need to develop smart tourism systems that adapt to the needs of users, enrich the tourist experience and increase the competitiveness of tourist destinations. Currently, there is a plethora of smart tourism and culture recommender systems that aim to offer personalised services to a tourist, improving the tourist experience overall [11].

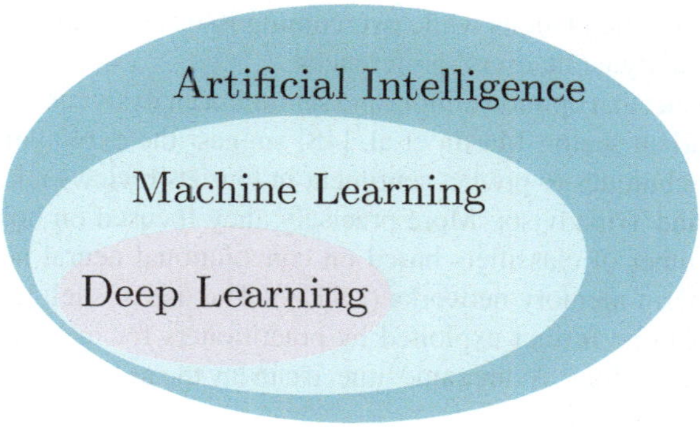

Fig. 7.4 Artificial intelligence, machine learning and deep learning [18]

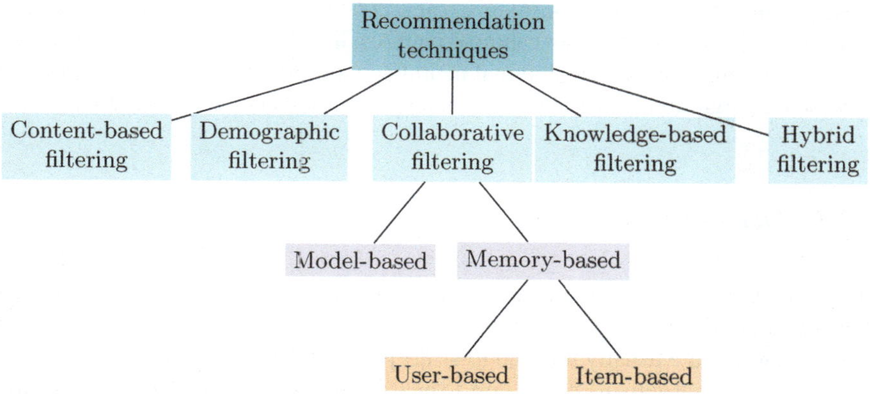

Fig. 7.5 Recommender system techniques

In the related literature, a vast number of papers focus on methods and categories of a recommender system, with the most common approaches being collaborative filtering, content-based filtering, demographic and knowledge-based recommender systems as well as hybrid ones, see Fig. 7.5. For more details regarding recommender systems, the interested reader may refer to [13, 15, 53, 63, 67].

As AI holds a crucial role in the creation of personalised offers [71], i during the last decade, we are witnessing a fundamental increase of research in artificial intelligence and deep learning recommender systems [46].As deep learning techniques empower image and object recognition, natural language processing and speech recognition, it comes as no surprise that researchers and practitioners exploit deep learning architectures in tourism recommendations process [7]. For example, authors in [66] exploit deep learning techniques to offer personalised point of interest (POI) recommendations. More specifically, they present a model they call Spatial-Aware Hierarchical Collaborative Deep Learning model that leverages users' check-in data and attempts to offer POIs recommendations while overcoming issues like data sparsity, the cold start problem, and dynamic users' preferences.

Based on the fact that opinion mining and sentiment analysis can be of great importance in the tourism sector, Martín et al. [48] suggest the exploitation of different deep learning techniques to predict sentiment in tourists reviews taken in particular form Booking and Tripadvisor. More precisely, they focused on hotel reviews and developed a number of classifiers based on convolutional neural networks (CNN) and long short-term memory networks (LSTM). The use of such techniques in the tourism sector can be further exploited by practitioners for marketing and service management applications. At the same time, it can lay the ground for further research in the smart tourism field.

Another research that exploits deep learning, and in particular Convolutional Neural Networks to realise personalised recommendations in the Smart Tourism sector is presented in [41]. The authors introduce a new concept, namely *Moments of Interest*, that derives from content of user captured pictures. In particular, the proposed crowdsourcing system utilises convolutional neural network for image labelling, face and sentiment recognition, and making personalised recommendations to tourists.

7.3.3 Deep Learning

Deep learning algorithms are considered a machine learning sub-domain [70]. As the plethora of deep learning techniques exploits neural network architectures, deep learning models are usually referred to as deep neural networks [49].

While the neural network term refers to neurobiology, deep learning models do not model humans' brain functions [18]. An artificial neural network refers to a parallel and distributed processor which consists of a number of independent computing units, namely neurons, that are interconnected to layers. An artificial neural network is exploited for a system to "learn" how inputs relate to outputs. In other words, the inputs and outputs are known, and the systems' goal is to "learn" the algorithm that "connects" them. On the contrary, in "traditional" programming the input and the algorithm that produces the output are known.

A neural network is a stack of layers, where each level consists of some mathematical function and internal variables ("weights" and "biases") that are updated during the training process. By training, we refer to the process that the neural network is repeatedly attempting to map an input to an output. All neurons in an artificial neural network are located in layers, where neurons in the same layer have the same activation function and behave in the same way. A neural network is composed of at least two layers, the input layer and the output layer. The interior layers are the so-called hidden layers.

7.3.4 Augmented Reality In tourism

Augmented reality is a rapidly growing research field, even though it exists for over five decades now [5, 14, 42]. The augmented reality term (AR) refers to the technology that "expands" the physical world with digital information, resulting in an enriched user experience. This "augmentation" of the real-world can be accomplished with data like videos, audio, 3D or 2D graphics as well as other computer-generated content [44]. Some key features of AR applications are the combination of the real world with the virtual, the real-time user interaction with the AR environment, as well as the alignment of the virtual objects with the real ones so as to create the illusion that these two worlds, the physical and the virtual, coexist harmoniously.

The aforementioned term is considered to have first appeared in 1990 by Tom Caudell and David Mizell. The AR concept, though, existed years before, when in the World War II, the British military developed a technology capable of displaying radar information on the windshield of a fighter plane [8]. Since then, many things have changed, and with the rise of the internet and smartphones, AR is now having an actual impact on users' habits, social life and the entertainment industry. At the same time, it is used in a plethora of sectors such as medicine, architecture, military training and education. The exploitation of AR tourism sector comes as no surprise. For example, Nóbrega et al. [51] propose a location adaptive augmented reality tourism system where, while in [58] Shih et al. exploit 3D scans in mobile an AR tourism application.

When developing an AR system, the key questions that arise are the type of digital content that will be displayed as well as where the content mentioned above is going to be displayed. In some cases, identification of specific physical objects is required, while in others it is just a location that is needed for the "augmentation" to take place. In general, augmented reality applications can be categorised into the following three categories:

- **Geo-located Augmented Reality applications**: In these applications, AR is triggered by users' geographical location obtained from sensors such as GPS and IMU. A quite popular location-based Augmented Reality application is Pokemon Go [50]. AR inspired or assisted guides could significantly improve the tourist experience. For instance, they could illustrate the former state of a hystoric building that has been damaged, present some context or people of previous ages to make the user feel as going back in time or even allow her to interact with the environment.
- **Marker-based Augmented Reality applications**: These applications require the detection of a marker for the augmentation to take place. This marker (target) can be a two-dimensional image, which consists of features that an algorithm can easily export, or a real-world object [17]. In other words, in such systems, there is a direct correlation between the real-world objects and the digital content. To exemplify, if an AR application is expected to display an animated graphic on a particular page of a book, then that application must first recognise the page via

camera (e.g. image/text recognition) and then display the digital content on it. Computer vision and image processing techniques are commonly used to locate markers.

- **Markerless augmented reality**: Markerless AR does not require the detection of an object for an augmentation to occur. It enables the exploitation of any or all parts of the natural environment as a target or basis for the display of digital virtual objects [62]. An example of such an AR system can be considered a home decoration application where the user places virtual furniture in his/her home. Techniques used to search for features in the scene are feature tracking and model-based tracking techniques, which are combined with pattern recognition methods.

7.3.5 AI Autonomous Agents

Artificial intelligence agents can be considered autonomous entities that exploit sensors and knowledge to serve a purpose and achieve a goal [43]. These agents that derive from the AI field are capable of carrying out operations on behalf of other users or systems independently [43]. Some main characteristics of this technology are its autonomy, reactivity, proactiveness, and social ability [43]. That being the case, it comes as no surprise that AI agents are exploited in a plethora of fields like health, weather forecast, education, and tourism [22].

More specifically, autonomous agents seem to also play a crucial role in the transformation of the tourism industry, and the realisation of an enhanced tourism experience [60]. There is a number of different approaches regarding the exploitation of such agents. An example would be text chatbots that assist travellers via replying to their messages. Other conversational agents may engage users via personalised recommendations, while others may handle tourists' feedback [22]. For instance, D'Haro et al. [22] present a conversational agent which can reply to natural language questions in the context of conference and tourism. KLM, the Dutch airline, also exploits text chatbot technology, namely Blue Bot that via deep learning techniques supports text base communication with customers [26]. A web agent-based intelligent recommendation application that aims at suggesting tour packages is presented in [33]. Garrido et al. [28] presented a system that tries to simulate a human guide. Their system utilises artificial intelligence techniques to extract and provide information to each user in a personalised manner. To achieve that, virtual characters that can provide interactive communication with people are utilised, namely Embodied Conversational Agents (ECA). The ultimate goal is to provide a way to interact directly with agents, in natural language, who can respond appropriately regarding tourist attractions and culture.

Voice assistants are also AI autonomous agents that may understand humans' requests and reply with synthetic voice [9]. An example of such voice-driven agents are Siri (Apple), Cortana (Microsoft), Alexa (Amazon), and Google's Assistant [19]. The rising interest in voice assistants in the tourism field can also be verified by the fact that Amazon is developing a particular version of Alexa just for the hospitality sector [9]. Connie, a Watson enabled robot developed in collaboration with IBM and deployed in Hilton, interacts with guests and recommends points of interest to them [35]. IBM Watson utilises deep learning techniques to create artificial neural networks and learn from unstructured data such as images, videos and texts. More specifically, IBM Watson uses transfer learning which is based on the exploitation of prior knowledge, so training from scratch is not required.

AI agents, since they first appeared, based on their design, can be considered as human-like and non-human-like [64]. Human-like service robots can be regarded as autonomous agents as based on the definition given by the international International Federation of Robotics [36], they can perform services with full or partial autonomy to humans [68]. As the idea of adopting anthropomorphic machines is growing to be more appealing worldwide, the tourism sector could not remain unaffected of this trend. Some examples would be Sacarino, a humanoid robot that assists hotel guests [69] and a trolley robot used to escort hotel customers and transfer their luggage [47]. Actually, the Henn-na Hotel in Nagasaki, Japan poses a fascinating example as it can be considered the worlds' first robot-staffed hotel, with robots "working" in the front desk and customer information points. Even though such a hotel may be tempting to visit, it must be pointed out that it actually reduced its robotic workforce by more than half, as it was not as efficient as expected [34]. Evidently, human-provided services are still an indispensable part of the hospitality sector.

7.4 Smart Tourism in COVID-19 Pandemic

The Coronavirus Pandemic undoubtedly turned the world upside down in so many ways. Out of a sudden, countries were forced into lockdowns while public health care systems around the world are continuously stressed to their limits. COVID-19 is a highly infectious virus and lethal in many cases, while in others, individuals affected may exhibit severe to mild symptoms or even be asymptomatic [21]. At the time of writing, the COVID-19 is affecting almost every country in the world, with struggling to contain its second wave. Global travel restrictions and limitation to all non-essential travel across borders was one of the measures to prevent the spread of the virus. As a result, the global tourism system went from over-tourism to non-tourism, with the usually overcrowded touristic sites all over the world being unprecedentedly deserted [30]. More specifically, based on statistics and insights provided by UNWTO (World Tourism Organization), that this COVID-19 crisis led to a loss in the scale of 850 million to 1.1 billion international tourists, which accounts for a loss of $910 billion to $ 1.2 trillion in export revenues from tourism, compromising 100 to 120 million jobs [61]. As the tourism industry has been one of the worst affected of this outbreak [16], it is of great importance for countries to be able to restart tourism and at the same time ensure travellers' well-being. Considering the above Gretzel et al. [31] issued a call for transformative research in the e-Tourism sector stating that: " *e-Tourism research must now elevate itself to be ready to ask, and prepare to find answers to, some critical and significant questions regarding the role of IT in contributing to a better future of tourism and society.*".

In the current pandemic COVID-19 crisis, many of the aforementioned AI approaches could adjust to assist with the virus containment while enabling tourism to restart. To exemplify, smart tourism recommender systems could exploit real-time users' location as well as neural networks predictions to propose less crowded places to tourists or specific time that points of interest can be visited to avoid interaction with many visitors. Moreover, AI autonomous agents technologies could restrict face-to-face human interaction, that in such difficult eras could pose a solution to the reduction of the coronavirus disease spread.

Moreover, in the worldwide literature, there have been already a number of research papers that attempt to exploit technological means to deal, in some way, with COVID-19 pandemic. To exemplify, authors in [2] present a survey of COVID-19 contact tracing apps that are exploited in some regions. The aforementioned systems attempt to automate the process of locating recent contacts of infected individuals. The plethora of these apps use a variety of technologies such as Global Positioning System, Bluetooth and Geographic Information System [16].

An example of such an application was Singapore's TraceTogether that uses Bluetooth technology to sense individuals within a specific distance. On the aforementioned app was based Australia's contact and trace app, known as COVIDSafe [1]. Such systems could also be useful in the context of smart tourism to realise the early tracing of COVID-19 infected individuals in travel destinations and eliminate both the spread of the virus and the effect this could have in the aforementioned destinations. As Ahmed et al. [2] noted, such systems may face issues related to users' privacy and security of collected data. At the same time, the trust of users to those accessing the mentioned above data is mandatory for the wide acceptance of contact tracing systems.

On the other hand, such techniques would be beneficial as they are cost effected and easily exploited in all tourism destinations, as they do not require expensive infrastructure [6]. Another approach to contain pandemic while resuming the international tourism sector is proposed in [3]. Angelopoulos et al. present a framework that shall enable the real-time sharing of information across multiple countries regarding whether an individual is infected with SARS-CoV-2. The aforementioned framework exploits a private blockchain and Proof of Authority for issuing Digital Health Passports.

A measure that could possibly contribute to public healthcare, as described by many governments, is to wear masks in public places like stores, public transportation even in the streets. Based on that, Jiang and Fan [37] propose a face mask detector, named RetinaFaceMask, that consists of ResNet or MobileNet as the backbone, feature pyramid network, and context attention modules. Moreover, it exploits transfer learning to adapt weights from networks trained for a similar face detection task. Another research paper that attempts to assist with this pandemic crisis via mask recognition is presented in [45]. The researchers show how a deep-learning-based artificial intelligence privacy-preserving mask detection artefact could look like, to ensure privacy (GDPR compliance) and high performance of mask recognition via a monitor. Such a system could be used for research purposes, to predict how mask coverage could assist pandemic containment or to monitor the actual compliance of store owners, museums or any other public places where the use of a mask is obligatory with this measure. Such a system in the tourism context could be exploited to assist both store owners and authorities in detecting violations related to wearing masks. In the case of adopting such a framework, though, extensive research is required to assure that human rights are not breached.

7.5 Conclusions and Future Directions

Artificial Intelligence is a field that continually draws researchers' and practitioners' attention in a plethora of sectors. While AI research seems highly promising, what is actually accomplished falls short of its expectancies [25]. In the smart tourism sector, while several researches focus on how the ICT revolution could alter the travel industry, still AI-related solutions are not as extensive as they could be [71]. Undoubtedly, AI impact on the tourism industry may only be described as beneficial [71]. Experimental results presented in the plethora of AI-related researches in this field only reinforce the aforementioned argument. Now more than ever, despite the severe blow to the tourism industry from the coronavirus, all technological means available need to be thrown into this unprecedented "war" to restart tourism.

Based on the above, the primary purpose of the present paper is to shed light on the role artificial intelligence holds in the smart tourism sector based on both published types of research as well as AI technologies exploited by stakeholders in the tourism industry. Based on the discussed and presented cases, AI can indeed lead to the realisation of a Smart Tourism experience. At the same time, we present technological proposals that could keep the tourism sector operational during the COVID-19 crisis period and potentially assist in restarting the travel and tourism activities.

From enhanced recommendation systems to service robots, AI can positively benefit the travel industry. At the time of writing, though, it is fair to say that AI in Smart Tourism is still at its infancy, with a lot of research gaps waiting to be filled. Firstly, even though recommender systems have already been widely exploited in the tourism context [12, 27, 29], still systems that successfully transform the vast amount of tourism-related heterogeneous data available, like images, videos, social media interactions and reviews, into knowledge are yet to be deployed. Future research should focus on exploiting such data in deep learning smart tourism recommender systems to achieve further personalised services, exactly tailored to the needs of the tourist and the context. Moreover, the need to progress from theoretical studies to the wide-scale exploitation of such systems in the tourism context is crucial for their evaluation and improvement. At the same time, the actual integration of AI smart tourism recommender systems could lead to the development of apps that exploit the "wisdom of the crowd" to upgrade the whole tourism experience.

Furthermore, artificially intelligent agents need to be further assessed and improved to attract and assist tourist without causing frustration for them. For example, systems that directly talk to travellers and actually understand them could make communication more natural and intimate, resulting in a better tourism experience. If we consider the Henn-na Hotel in Nagasaki that ended up in reducing its robotic workforce to more than half, it is clear that autonomous agents are far from being as efficient as expected in literature. Finally, in these challenging times, we are living, a new way of traveling is required. There is an imminent need for the tourist industry to adjust to this new reality and exploit technology to make travelling safe again. Thus, technologies mentioned in Sect. 7.4 should be taken into consideration and be adapted to the tourism industry's needs. However, concerns regarding the social acceptance of technological approaches along with the need to always comply with human rights and privacy-preserving regulations need to be seriously taken into consideration for research proposals to be implemented in the real world.

References

1. R. Abbas, K. Michael, Covid-19 contact trace app deployments: Learnings from australia and singapore. IEEE Consumer Electron. Mag. **9**(5), 65–70 (2020)
2. N. Ahmed, R.A. Michelin, W. Xue, S. Ruj, R. Malaney, S.S. Kanhere, A. Seneviratne, W. Hu, H. Janicke, S.K. Jha, A survey of covid-19 contact tracing apps. IEEE Access **8**, 134577–134601 (2020)
3. C.M. Angelopoulos, V. Katos, Dhp framework: Digital health passports using blockchain-use case on international tourism during the covid-19 pandemic (2020)
4. O. Artemenko, V. Pasichnyk, H. Korz, P. Fedorka, Y. Kis, Using big data in e-tourism mobile recommender systems: a project approach. in *Proceedings of the 1st International Workshop IT Project Management (ITPM 2020)*, eds. by S. Bushuyev, V. Lytvyn, N. Kunanets, V. Pasichnyk (Slavsko, Lviv region, Ukraine, 2020). CEUR Workshop Proceedings, vol. 2565, pp. 194–204. CEUR-WS.org (2020). http://ceur-ws.org/Vol-2565/paper17.pdf
5. R.T. Azuma, A survey of augmented reality. Teleoperators & Virtual Environments, Presence **6**(4), 355–385 (1997)
6. A. Bansal, R.P. Padappayil, C. Garg, A. Singal, M. Gupta, A. Klein, Utility of artificial intelligence amidst the covid 19 pandemic: a review. J. Med. Syst. **44**(9), 1–6 (2020)
7. Z. Batmaz, A. Yurekli, A. Bilge, C. Kaleli, A review on deep learning for recommender systems: challenges and remedies. Artif. Intell. Rev. **52**(1), 1–37 (2019)
8. D.R. Berryman, Augmented reality: a review. Med. Ref. Serv. Q. **31**(2), 212–218 (2012)
9. T. Bittendorfer, J. Bunt, L. Grunder, D. Riedel, B. Magnus, T. Salzlecher, Technology in tourism: How voice assistants influence the hospitality industry. in *ISCONTOUR*, pp. 328–338 (2019)
10. T. Brandt, J. Bendler, D. Neumann, Social media analytics and value creation in urban smart tourism ecosystems. Inf. Manag. **54**(6), 703–713 (2017)
11. M. Braunhofer, F. Ricci, Selective contextual information acquisition in travel recommender systems. J. IT Tourism **17**(1), 5–29 (2017). https://doi.org/10.1007/s40558-017-0075-6
12. M. Braunhofer, F. Ricci, Selective contextual information acquisition in travel recommender systems. Inf. Technol. Tourism **17**(1), 5–29 (2017). https://doi.org/10.1007/s40558-017-0075-6
13. R. Burke, Hybrid web recommender systems. in *The adaptive web* (Springer, Berlin, 2007)
14. J. Carmigniani, B. Furht, M. Anisetti, P. Ceravolo, E. Damiani, M. Ivkovic, Augmented reality technologies, systems and applications. Multimed. Tools Appl. **51**(1), 341–377 (2011)

15. F. Casino, C. Patsakis, D. Puig, A. Solanas, On privacy preserving collaborative filtering: Current trends, open problems, and new issues, in *2013 IEEE 10th International Conference on e-Business Engineering* (IEEE, 2013), pp. 244–249
16. V. Chamola, V. Hassija, V. Gupta, M. Guizani, A comprehensive review of the covid-19 pandemic and the role of iot, drones, ai, blockchain, and 5g in managing its impact. IEEE Access **8**, 90225–90265 (2020)
17. J. Cheng, K. Chen, W. Chen, Comparison of marker-based ar and marker-less ar: a case study on indoor decoration system, in *Lean and Computing in Construction Congress (LC3): Proceedings of the Joint Conference on Computing in Construction (JC3)* (2017), pp. 483–490
18. F. Chollet, *Deep Learning with Python*, 2nd edn. (Manning Publications, 2020)
19. R. Dale, The return of the chatbots. Nat. Lang. Eng. **22**(5), 811–817 (2016)
20. DBLP: Dblp (2019). Available at https://dblp.uni-trier.de/, visited on August 1, 2019
21. Of Disease control, C., Prevention: Coronavirus disease 2019 (covid-19) (2020), available at https://www.cdc.gov/coronavirus/2019-ncov/symptoms-testing/symptoms.html, visited on November 6, 2020
22. L.F. D'Haro, S. Kim, K.H. Yeo, R. Jiang, A.I. Niculescu, R.E. Banchs, H. Li, Clara: a multifunctional virtual agent for conference support and touristic information, in *Natural Language Dialog Systems and Intelligent Assistants* (Springer, Berlin, 2015), pp. 233–239
23. P. Fettke, Conceptual modelling and artificial intelligence: Overview and research challenges from the perspective of predictive business process management, in *Modellierung* (2020)
24. M. Figueredo, N. Cacho, A. Thome, A. Cacho, F. Lopes, M. Araujo, Using social media photos to identify tourism preferences in smart tourism destination, in *2017 IEEE International Conference on Big Data (Big Data)* (2017), pp. 4068–4073. https://doi.org/10.1109/BigData.2017.8258423
25. R. Fjelland, Why general artificial intelligence will not be realized. Human. Soc. Sci. Commun. **7**(1), 1–9 (2020)
26. T. Gajdošík, M. Marciš, Artificial intelligence tools for smart tourism development, in *Computer Science On-line Conference* (Springer, Berlin, 2019), pp. 392–402
27. L.M. García, S. Aciar, R. Mendoza, J.J. Puello, Smart tourism platform based on microservice architecture and recommender services, in *Mobile Web and Intelligent Information Systems—15th International Conference, MobiWIS 2018, Proceedings* (Barcelona, Spain, 2018) (2018), pp. 167–180. https://doi.org/10.1007/978-3-319-97163-6_14
28. P. Garrido, J. Barrachina, F.J. Martinez, F.J. Serón, Smart tourist information points by combining agents, semantics and AI techniques. Comput. Sci. Inf. Syst. **14**(1), 1–23 (2017)
29. D. Gavalas, C. Konstantopoulos, K. Mastakas, G. Pantziou, Mobile recommender systems in tourism. J. Netw. Comput. Appl. **39**, 319–333 (2014). https://doi.org/10.1016/j.jnca.2013.04.006
30. S. Gössling, D. Scott, C.M. Hall, Pandemics, tourism and global change: a rapid assessment of covid-19. J. Sustain. Tourism 1–20 (2020)
31. U. Gretzel, M. Fuchs, R. Baggio, W. Hoepken, R. Law, J. Neidhardt, J. Pesonen, M. Zanker, Z. Xiang, e-tourism beyond covid-19: a call for transformative research. Inf. Technol. Tourism 1 (2020)
32. U. Gretzel, M. Sigala, Z. Xiang, C. Koo, Smart tourism: foundations and developments. Electron. Markets **25**(3), 179–188 (2015). https://doi.org/10.1007/s12525-015-0196-8
33. R. Hassannia, A. Vatankhah Barenji, Z. Li, H. Alipour, Web-based recommendation system for smart tourism: Multiagent technology. Sustainability **11**(2), 323 (2019)
34. E. Hertzfeld, Japan's henn na hotel fires half its robot workforce (2019). Available at https://www.wired.com/2016/03/robot-henn-na-hotel-japan/, visited on November 2, 2020
35. IBM: Watson connie (2016). Available at https://www.ibm.com/blogs/watson/2016/03/watson-connie/, visited on November 2, 2020
36. IFR: International federation of robotics (2020). Available at https://ifr.org/, visited on November 2, 2020
37. M. Jiang, X. Fan, Retinamask: A face mask detector. arXiv:2005.03950 (2020)

38. L.P. Kaelbling, M.L. Littman, A.W. Moore, Reinforcement learning: A survey. J. Artif. Intell. Res. **4**, 237–285 (1996)
39. M.K. Khan, Technological advancements and 2020. Telecommun. Syst. **73**(1), 1–2 (2020). https://doi.org/10.1007/s11235-019-00647-8
40. J.N. Kok, E. Boers, W.A. Kosters, P. Van der Putten, M. Poel, Artificial intelligence: definition, trends, techniques, and cases. Artif. Intell. **1**, 1–20 (2009)
41. A. Kontogianni, E. Alepis, Moments of interest: A novel cloud-based crowdsourcing application enhancing smart tourism recommendations, in *2019 11th Computer Science and Electronic Engineering (CEEC)* (IEEE, 2019), pp. 144–149
42. A. Kontogianni, E. Alepis, Smart tourism: State of the art and literature review for the last six years. Array **6**, 100020 (2020)
43. M.K. Kouluri, R. Pandey, Intelligent agent based micro grid control, in *2011 2nd International Conference on Intelligent Agent and Multi-Agent Systems* (IEEE, 2011), pp. 62–66
44. C.D. Kounavis, A.E. Kasimati, E.D. Zamani, Enhancing the tourism experience through mobile augmented reality: Challenges and prospects. International Journal of Engineering Business Management **4**, 10 (2012). https://doi.org/10.5772/51644
45. N. Kühl, D. Martin, C. Wolff, M. Volkamer, Healthy surveillance: Designing a concept for privacy-preserving mask recognition ai in the age of pandemics [in press], in *Hawaii International Conference on System Sciences (HICSS 54)*, online (2020)
46. K.H. Lai, N.Y. Yen, M.Y. Chen, Design of an ai-empowered recommender system for travelling support: Individual traveler as an instance, in *2018 IEEE 16th International Conference on Dependable, Autonomic and Secure Computing, 16th International Conference on Pervasive Intelligence and Computing, 4th International Conference on Big Data Intelligence and Computing and Cyber Science and Technology Congress (DASC/PiCom/DataCom/CyberSciTech)* (IEEE, 2018), pp. 343–346
47. G. Lewis-Kraus, *Check in With the Velociraptor at the World's First Robot Hotel*. Wired Magazine (2016)
48. C.A. Martín, J.M. Torres, R.M. Aguilar, S. Diaz, Using deep learning to predict sentiments: case study in tourism. Complexity **2018** (2018)
49. MathWorks: What is deep learning? (2020). Available at https://www.mathworks.com/discovery/deep-learning.html, visited on October 23, 2020
50. Nintendo: Pokemon go (2020). Available at https://pokemongolive.com/en/, visited on October 29, 2020
51. R. Nóbrega, J. Jacob, A. Coelho, J. Ribeiro, J. Weber, S. Ferreira, Leveraging pervasive games for tourism: An augmented reality perspective. IJCICG **9**(1), 1–14 (2018). https://doi.org/10.4018/IJCICG.2018010101
52. Researchgate: Researchgate (2019). Available at https://www.researchgate.net/, visited on August 7, 2019
53. F. Ricci, L. Rokach, B. Shapira, Recommender systems: introduction and challenges. in *Recommender Systems Handbook* (Springer, Berlin, 2015), pp. 1–34
54. S. Russell, P. Norvig, *Artificial Intelligence: A Modern Approach* (Prentice Hall, 2002)
55. G. Scholar, Googlescholar (2019). Available at https://scholar.google.gr/, visited on August 7, 2019
56. ScienceDirect: Artificial intelligence (2020). Available at https://www.sciencedirect.com/topics/computer-science/artificial-intelligence, visited on November 7, 2020
57. Scopus: Scopus (2019). Available at https://www.scopus.com/home.uri, visited on August 7, 2019
58. N. Shih, P. Diao, Y. Chen, Arts, an AR tourism system, for the integration of 3d scanning and smartphone AR in cultural heritage tourism and pedagogy. Sensors **19**(17), 3725 (2019)
59. Statista: Global tourism industry—statistics & facts (2020). Available at https://www.statista.com/topics/962/global-tourism/, visited on November 7, 2020
60. N. Stylos, Technological evolution and tourist decision-making: a perspective article. Tourism Rev. (2019)

61. UNWTO: Unwto tourism data dashboard (2020). Available at https://www.unwto.org/unwto-tourism-dashboard, visited on November 6, 2020
62. A.F. Wartawu, I.P.A. Bayupati, I.K.G.D. Putra, Augmented reality mobile application of balinese hindu temples: Dewataar. Int. J. Comput. Netw. Inf. Secur. **7**(2), 59 (2015)
63. J. Wei, J. He, K. Chen, Y. Zhou, Z. Tang, Collaborative filtering and deep learning based recommendation system for cold start items. Exp. Syst. Appl. **69**, 29–39 (2017)
64. J. Wirtz, P.G. Patterson, W.H. Kunz, T. Gruber, V.N. Lu, S. Paluch, A. Martins, Brave new world: service robots in the frontline. J. Serv. Manag. (2018)
65. B.H. Ye, H. Ye, R. Law, Systematic review of smart tourism research. Sustainability **12**(8), 3401 (2020)
66. H. Yin, W. Wang, H. Wang, L. Chen, X. Zhou, Spatial-aware hierarchical collaborative deep learning for poi recommendation. IEEE Trans. Knowl. Data Eng. **29**(11), 2537–2551 (2017). https://doi.org/10.1109/TKDE.2017.2741484
67. P. Yochum, L. Chang, T. Gu, M. Zhu, Linked open data in location-based recommendation system on tourism domain: A survey. IEEE Access **8**, 16409–16439 (2020)
68. C.E. Yu, H.F.B. Ngan, The power of head tilts: Gender and cultural differences of perceived human vs human-like robot smile in service. Tourism Rev. (2019)
69. E. Zalama, J.G. García-Bermejo, S. Marcos, S. Domínguez, R. Feliz, R. Pinillos, J. López, Sacarino, a service robot in a hotel environment, in *ROBOT2013: First Iberian Robotics Conference* (Springer, Berlin, 2014), pp. 3–14
70. S. Zhang, L. Yao, A. Sun, Y. Tay, Deep learning based recommender system: A survey and new perspectives. ACM Comput. Surv. (CSUR) **52**(1), 1–38 (2019)
71. M. Zsarnoczky et al., How does artificial intelligence affect the tourism industry? VADYBA **31**(2), 85–90 (2017)

Chapter 8
Challenges and AI-Based Solutions for Smart Energy Consumption in Smart Cities

Miltiadis Alamaniotis

Abstract In the smart cities of the future artificial intelligence is expected to play an important role in facilitating residents' lives while widening the variety of services offered to them. This role will focus on collecting, analyzing data, and making projections of the needs and availability of resources of the city in short terms. In a broader view, the term "smart city" encompasses the coupling of the current infrastructure with information networks and the dissipation of information to people within the city limits in real-time. The information provided by the city infrastructure to the residents via several communication channels will allow them to make optimal decisions concerning their energy consumption. The optimal management of energy systems via information and data utilization is known with the term "smart energy". AI-driven solutions can accommodate the realization of smart energy systems by optimizing consumption at the level of a single resident, of a neighborhood, or a cluster of residents. This chapter will discuss challenges and AI-driven technological solutions pertained to smart energy services within the context of a smart city.

8.1 Introduction

The advancements in data science and technologies have fostered the ground for realizing the vision of smart cities [1]. This vision encompasses the coupling of the current city infrastructure with information networks, thus accommodating the digital connectivity of the city with its residents [2]. Digital connectivity will allow the deployment of sensor networks within the city premises and the unification of the city components into a single entity that interfaces with the residents [3]. It is expected, that smart cities will consist of an information-rich environment, where utilization of information promotes the optimal city operation resulting in a clean, friendly, and green environment leveraging the quality of life for humans [4–6].

M. Alamaniotis (✉)
Department of Electrical and Computer Engineering, The University of Texas at San Antonio, 1 UTSA Circle, San Antonio, TX 78249, USA
e-mail: miltos.alamaniotis@utsa.edu

M. Virvou et al. (eds.), *Advances in Artificial Intelligence-based Technologies*, Learning and Analytics in Intelligent Systems 22,
https://doi.org/10.1007/978-3-030-80571-5_8

From a system point of view, smart cities may be seen as the synthesis of several components that are depicted in Fig. 8.1. The fundamental component is the "digital connectivity" that takes the form of data networks and allows the realization of the vision of a smart city. Furthermore, essential is the transformation of citizens to "digital citizens" (as shown in Fig. 8.1), via the adoption of digital systems (for instance, IoT devices and cell phones [7]) that accommodate the connection of citizens to the rest of the smart city components. "Open data" is fundamental for the smooth and efficient operation of the smart city given that it will provide citizens with access to data useful for making optimal decisions. The rest of the components refer to services accessible to citizens pertained to their daily needs and habits. These services are identified as "smart", a term that implies the use of information technologies to optimize their operation and increase their availability [8]. Namely, the set of smart services encompasses: "smart manufacturing", "smart government", "smart health", "smart agriculture", "smart buildings", "smart energy" and "smart transportation".

Of special interest in the operation of the city is the smart energy and grid component. Smart energy, mainly perceived in the form of electrical energy, is essential for retaining and the current quality of life [9]. Besides, it is the cornerstone upon which all the rest of smart services are being built: energy is the driving force behind the use and availability of the rest of the city-related services. For instance, the pyramidic visualization of the city services shown in Fig. 8.2 places the "smart energy/grid" at the bottom of the structure representing the most important piece of the city services [10]. Our society is built upon the extensive use of electrical energy and therefore its 24/7 availability and utilization are mandatory for retaining this type of life; shortage of electricity for while causes chaos and panic as has been observed in small-scale blackouts cases [11]. Analogously, smart energy secures the availability and utilization of the rest of the smart services. For instance, smart energy systems secure

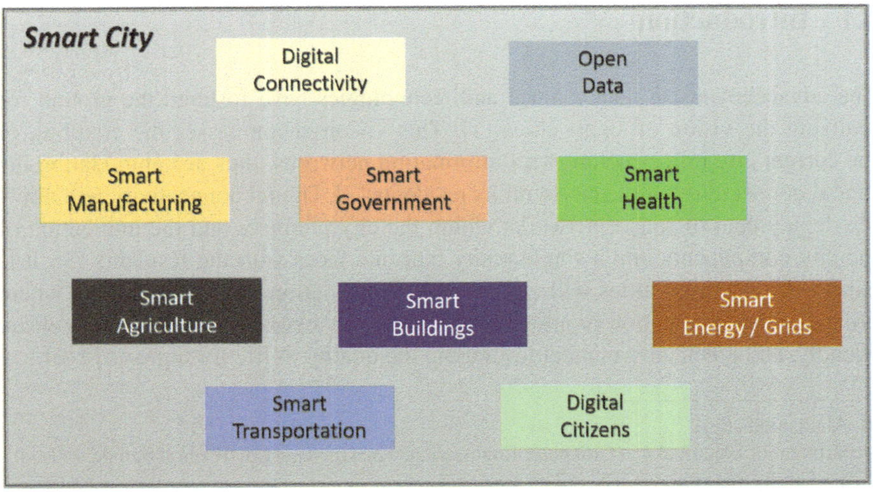

Fig. 8.1 Smart city components

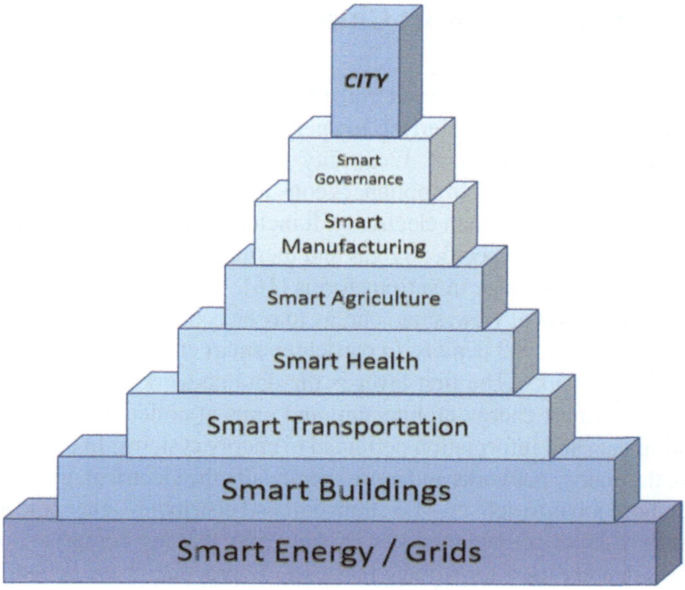

Fig. 8.2 Pyramidic representation of smart city services with smart energy/grids being the most important part of it

the resilient operation of the grid and therefore, secure the electricity flow which is required for implementing the smart transportation and smart health systems; a problem in energy flow will inevitably impact the operation of the other services [12].

In this chapter, the smart energy challenges within the frame of smart cities will be presented and the role of artificial intelligence solutions will be discussed. The main goal of smart energy is to optimize the consumption of energy in the city ensuring the continuous operation of the energy grid [13] and attain a significant reduction of carbon emissions [14]. Notably, smart energy is dominated by the use of electricity consumption though it includes other types such as gas, renewable energy that is utilized directly and is not converted to electrical energy [15].

The roadmap of this chapter is as follows. The next section introduces the concept of smart energy in smart cities. Section 3 discusses the challenges and the AI solutions in energy consumption, while Sect. 4 concludes and summarizes the main points of the chapter.

8.2 Smart Energy in Smart Cities

The term of smart energy has emerged from the coupling of information technologies with energy networks. When referring to energy the dominant form of energy that comes to our minds is electrical. Electricity is the driving force behind modern civilization given that all modern appliances consume electricity to operate. However, smart energy contains more than electricity. It includes all the sources of energy that may be utilized and more specifically gas and thermal energy, while it also contains the concept of energy storage in various forms [16].

From a broader point of view, smart energy may be considered as the integration of several networks as Fig. 8.3 depicts. In particular, smart energy architecture contains three layers of networks. The first layer is the data network which consists of the backbone of the smart energy architecture and aims at collecting, processing, and disseminating data and information pertained to energy systems. In the second layer, we observe the energy networks and more specifically the electrical, gas, and thermal networks, whose job is to deliver—i.e., transport and distribute—energy to end-users. Lastly, the third layer contains the storage networks that are comprised of devices and facilities that are used for storage purposes. In practice, these networks are not physically interconnected but they may be considered as such; in practice, some of them are connected to the energy networks: for instance, batteries are connected to the electricity grid. Notably, in Fig. 8.3, we observe that the group of storage networks encompasses four energy storage components, namely, gas, thermal, battery, and liquid fuel storage.

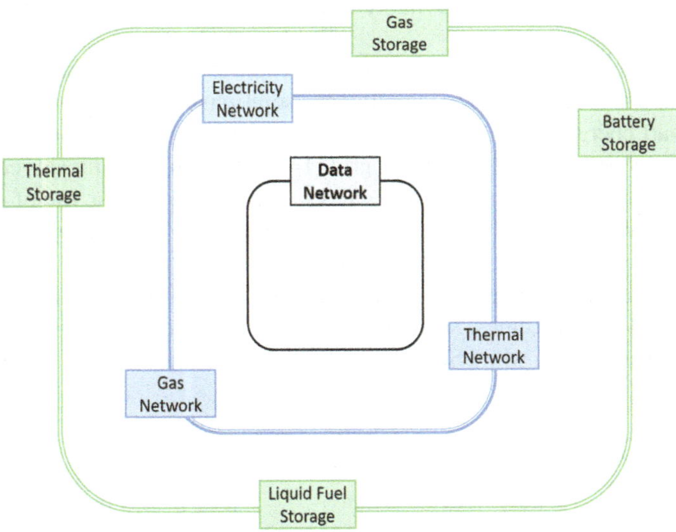

Fig. 8.3 Representation of network architecture involved in smart energy systems

It is needless to say, that if the aforementioned networks are located within the premises of a city, then they are part of the smart city infrastructure. Smart energy may provide a diverse portfolio of energy sources for the smart city residents who can make optimal decisions for their needs. For instance, city residents participating in a price-directed energy market may optimize the cost of heating their house by combining purchases of gas and electricity, thus, creating an optimal mixed energy portfolio.

Energy storages within the cities are considered essential for implementing smart energy systems. It should be noted that storage devices may belong to residents, like batteries, or maybe part of the public infrastructure—e.g., fuel storage containers located in the port of a city.

One of the main challenges in smart energy is the satisfaction of the demand of people of different statuses and lifestyles in close geographic proximity. For instance, residents in wealthy areas may have higher demand and a high variety of ways for energy utilization. To make it clearer, for instance, luxurious areas may have gas pipes for supply, while they may have their own electricity generators [17]. On the other hand, poor areas may utilize only electricity, while their demand is low. Also, wealthy residents may afford and adopt a high number of technological advancements—e.g., smart home technologies-, while poor residents are less prone to use new technologies (because of the purchasing cost) [18]. Therefore, those differences make the city regions exhibit differences and the overall city heterogeneous concerning energy consumption.

Driven by those differences, it can be stated that the city can be partitioned into sectors based on the average consumption profile. Thus, it can be said without loss of generality that a smart city may be portioned into various consumption sectors [19], with each sector exhibiting a consumption profile that characterizes the wealth and lifestyle of the residents in that sector [20].

Overall, energy consumption is an essential issue in the normal operation of the cities. Consequently, the leverage of cities to smart cities will impose several challenges to smart energy systems concerning residents' consumption. However, smart cities have a powerful ally: the utilization of information. Therefore, challenges and constraints that are inherent in the physical energy networks can be overcome by the intelligent use of information. The word "intelligent use" is strongly associated with the use of artificial intelligence tools [21].

8.3 Energy Consumption Challenges and AI Solutions

8.3.1 End-User Consumers in Smart Cities

The goal of energy and subsequently of power systems is the generation and distribution of energy to end-users. In a smart city, the end-users may be identified as one of the following types: (i) residential, (ii) commercial, (iii) industrial, (iv) public infrastructure, and (v) mobile end users. In particular, the characteristics of each type are given below:

- *Residential end-users*: This group contains energy consumers that are identified as residents containing houses and apartments. In other words, it contains the residential buildings within a city where people spent their time. These end-users mainly utilize energy for activities that have to do with the daily needs of people (cooking, sleeping, washing, watching tv, etc.). The profile of residents exhibits high variance and volatility, and it is the most common type of end-user found in smart cities.
- *Commercial end-users*: This group contains business and small commercial buildings. The main characteristic of this group is that they exhibit high consumption during business hours, while they have low consumption at night and weekends. Besides, consumption during the business exhibits low volatility and large rises and drops at the beginning and end of business hours. Similar to residential end-users, commercial users are also commonly found in smart cities.
- *Industrial end-users*: This is one of the less common types of end-users found in smart cities. It is usual that industries have large energy demands and are located outside of the city limits. However, there are some cases where industrial end-users lie within the smart city limits and therefore their consumption is handled as part of the overall smart energy portfolio. They are the less inelastic consumers compared to the aforementioned ones while having large demands, especially during business hours. Several industrial operate on weekends and thus they require large amounts of energy even during those times.
- *Public infrastructure end-users*: The backbone of a city is its infrastructure. Likewise, the backbone of a city is its infrastructure with the information network being the cornerstone. Operation of the infrastructure requires the consumption of energy with the most common form being electricity. Therefore, the public infrastructure has its energy demand (i.e., for continuous operation) and is identified as an end-user.
- Mobile end users*:* This is a special category and refers to all energy consumers that are mobile within the limits of the city. The use of information technologies and the connection of the residents to the public infrastructure requires the use of devices (e.g., cell phones, tablets, car monitors, etc.). Also, vehicles that move within the limits of the city—either electrical or gas-fueled—need to connect to the infrastructure of the city for energy supplies. Therefore, they consist of a type of end-users with a special characteristic that is not located in a permanent position but is mobile. Furthermore, cars based on liquid fuel (gas) are also considered to fall within this type of end-users. It should be noted that with energy remote transfer technologies that are under development, this type of end-users will become very popular in the future. Remote charging of devices via the smart city infrastructure will only increase the demand for this type of end-users [22].

The above classification of end-users in smart cities imposes several challenges in energy consumption. The main challenges contain demand forecasting, prosumers management, and consumption privacy.

8.3.2 *Demand Forecasting*

8.3.2.1 Challenge

One of the main problems since humanity started utilizing energy sources was the forecasting of the demand at every time instance. Forecasting—in other words foreseeing the amount of energy that will be needed—is essential for the efficient and optimal operation of the energy networks and subsequently for the city [23].

In particular demand, forecasting accommodates the matching of the demand with the supply. Knowing the exact amount of demand at every time point prevents offers a group of benefits. First of all, shortcomings in energy supply are not possible given that forecasting allows the ahead of time scheduling of unit commitments to operate and produce the necessary energy. This refers to electricity generation units such as power plants (conventional and nuclear). Also, energy demand forecasting accommodates the secure operation of the energy networks given that by forecasting operators can foresee faults and failures of the energy networks.

The grand challenge of energy demand forecasting has several implications for the utility operators as depicted in Fig. 8.4. Energy forecasting allows:

- *Energy network failure/fault prediction*: the forecasting of energy demand allows the ahead of time prediction of possible faults or failures in one of the energy networks. Failures/faults mainly refer to the inability of the system to deliver the energy to the smart city end-user and therefore provide room for designing an alternative plan for energy delivery.
- *Supply shortage prediction*: In this case, the forecasting methods may predict whether the generated energy will be lower than the actual demand. Thus, it allows avoiding dissatisfaction with residents' needs by providing a mixed energy portfolio schedule to satisfy the needs. To make clear, forecasting contributes to customer satisfaction.

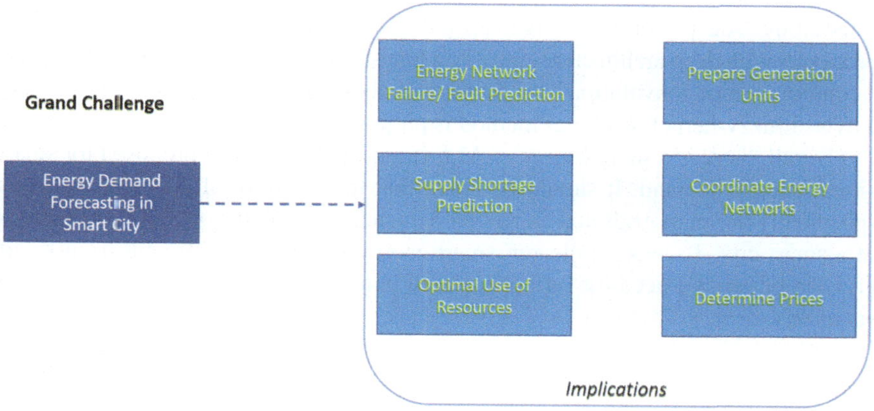

Fig. 8.4 Energy demand forecasting challenge and implications for utility operators in smart cities

- *Optimal use of resources*: In this case, the forecasting allows the optimal use of energy resources and the delivery means. Knowing exactly the demand allows the scheduling of energy delivery by identifying the ways that are of low cost, fast, and put less pressure on the infrastructure, thus, maintenance costs are minimized.
- *Prepare Generation units*: Accurate forecasting allows the preparation of the generation units given that some of them need startup time before generating, while it allows the scheduling of generation units according to their operational cost (i.e., lower-cost units are selected first).
- *Coordinate energy networks*: The utility operator requires forecasting so he/she can make optimal decisions about the allocation of the energy networks. The optimal decision will combine the energy networks in such a way that the utilization of energy is maximum and the cost is the minimum possible. For instance, gas networks can be combined with electrical grids to provide heat at low costs.
- *Determination of prices*: The demand is associated with the supply and inherently with the energy prices. Knowing a priori the energy needs, the operator can make an optimal energy portfolio to satisfy the demand. This scheduling is also the driving force behind the determination of the energy prices given the strong correlation between demand and prices [24].

8.3.2.2 AI Solutions in Forecasting

Artificial intelligence is one of the main springboards to provide effective solutions for energy demand in smart cities. Due to the complexity of the challenge and its implications as presented in the previous section, AI is considered to be an ideal toolbox for smart cities [25].

In particular, AI provides data-driven methods to provide forecasting given that it is being utilized to develop intelligent analytics methods. Those methods utilize historical patterns from the city residents and learn their behavior. Driven by the learned patterns the AI methods may provide future patterns of demand. In that direction several methods have been proposed and applied in smart city energy demand forecasting: a deep neural network in [26], the combination of kernel-based Gaussian processes with deep architectures in [27], a long-term short memory network in [28], a combination of convolutional networks with long term short memory in [28], and a hybrid fuzzy-kernel machines method is proposed in [29].

Overall, there is a short variety of AI methods that have been proposed for smart cities' energy demand. It should be noted that the majority of AI methods focus specifically on smart grids and electricity demand, ignoring the overall energy needs of a smart city. To make it clearer, smart energy systems are different from smart grids and there is a need for further research in the area of energy demand within a smart city.

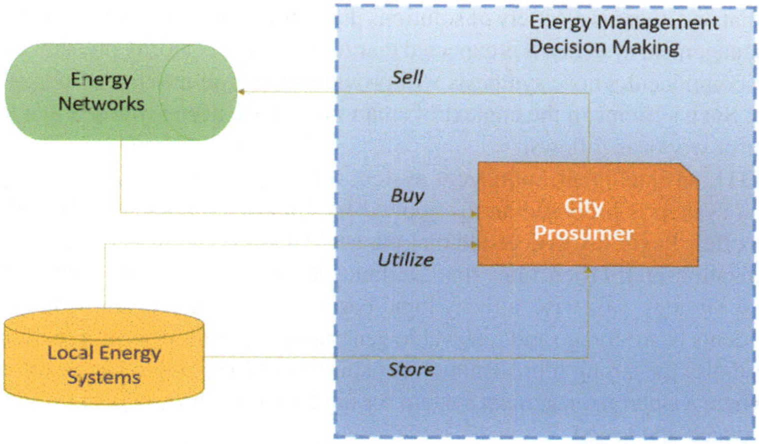

Fig. 8.5 Prosumer energy management challenge in smart cities

8.3.3 Prosumers Management

8.3.3.1 Challenge

The advancement in energy and power technologies via data tools, especially those of renewable sources, has transformed the typical consumers into a hybrid model of prosumers. The term "prosumers" refers to the ability of consumers not only to consume but also to produce and sell energy. This selling will be very common in smart cities. The advanced energy infrastructure and the high-tech delivery means together with the high population density in the smart cities accommodates the transformation of consumers to prosumers [30].

The main challenge in prosumers has to do with the management of energy and the decisions that have to be made. Prosumers have to make decisions on the amount of energy that needs to be bought, from which utility, and what type of energy will be purchased. Additionally, they have to decide how much of their energy will be utilized by themselves, where they will sell energy and the amount of selling. In other words, prosumers make two-directional decisions: buy and sell as shown in Fig. 8.5. However, these decisions must be combined to maximize the benefit of the prosumer concerning cost and demand satisfaction.

8.3.3.2 AI Solutions in Prosumer Management

Tools from the AI realm are utilized for implementing autonomous decision-making systems. Concerning prosumers, AI systems may be used for making energy management decisions, where management contains both purchasing and selling energy decisions.

To that end, there is a variety of solutions that utilize AI tools to implement intelligent management systems. It is expected that AI at the level of a city prosumer will be a basic component whose synthesis will provide an overall intelligent management system. Such systems in the context of smart cities have been proposed in a limited number of works as follows.

In [31], an intelligent multiagent system that implements a fuzzy anticipatory decision system is proposed and is applied for smart city residents. The proposed system offers flexibility in decision making while it allows a 2-way decision making via utilization of IF/Then rules that facilitate fast decision making and update of purchase strategy. Likewise, a fuzzy logic-based energy management system where the residents have storage and renewable generation capabilities is proposed in [32]. Furthermore, the strong representation capabilities of fuzzy logic were exploited in [33], where a smart management system for implementing prosumers in smart cities is introduced and tested.

In [34], Gaussian processes were utilized for load and price forecasting and their subsequent integration to provide an optimal schedule for energy management. The proposed method mainly focuses on electricity prices but by taking the average of prices and demands of each energy source respectively, the method can be utilized for prosumers.

In [35], an intelligent framework for efficient energy management in smart cities that applies to city prosumers is proposed. The proposed framework is based on a combination of deep learning and linear optimization.

Overall, AI has a strong potential for implementing autonomous management systems within the frame of city prosumers. This is a research area with a lot of potential that needs further exploration. So far, the majority of the methods are seen from a smart grid point of view, but this is a very limited approach if put in the context of smart cities.

8.3.4 Consumption Privacy

8.3.4.1 Challenge

One of the main challenges in smart cities is end-user privacy. Smart cities are expected to be information-rich environments where the exchange of information imposes privacy challenges. Concerning smart energy, consumption patterns are also a source of information that may be utilized for making inferences pertained to end-user actions, private life, and other aspects as shown in Fig. 8.6.

At this point, it should be noted that this chapter does not refer to privacy issues referred to data encryption. This is an issue that has been resolved by cybersecurity methods. In smart cities, there is the issue of the "third observer". For instance, someone that has access to consumption data and can sell the data to interested parts (for example a utility worker selling the consumption data to a marketing company). Unfortunately, there are a lot of third parties that have to monitor the consumption

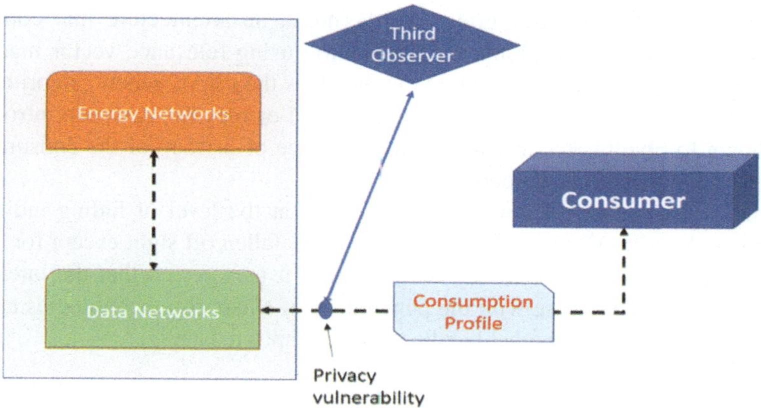

Fig. 8.6 Representation of privacy challenges in energy consumption in smart cities

profiles without any encryption code. An example is that someone who wants to purchase electricity has to send his schedule to the market operator; in case of an electricity market with bilateral contracts, the operators' staff have to observe the pattern and provide a price to the consumer to agree on, and thus there is no security that the consumption won't be secretly shared with third observers.

8.3.4.2 AI Solutions in Consumption Privacy

Research on encrypting data is an old science that goes back to the invention of written language. Therefore, there is a huge variety of encrypting data and sending it over a communication channel. However, in smart cities, the challenge is how to secure that the consumption data are not shared among interested parties. In this chapter, the interest is on how the consumption data can hide the individual consumption features of the city residents (the end-users defined earlier).

From that point of view, there is not much research that involves artificial intelligence; however, the limited number of works pave the way for more advanced methods using AI. To that end, the following methods have been introduced that manipulate the consumption pattern to enhance the privacy of consumers within smart cities.

In [36], a method that utilizes genetic algorithms and self-elasticities to enhance privacy in a group of residents within a smart city. The goal of the method is that to aggregate the consumption patterns of a specific group of residents and reschedule their consumption in such a way as to hide the individual characteristics.

Another method that used machine learning and more specifically Gaussian processes to anticipate the consumption patterns and subsequently morph via selective aggregation is introduced in [37]. Similarly, a particle swarm optimization-based method was proposed in [38] for morphing the aggregated energy demand response of a predetermined end-user group. Furthermore, an architecture that combines IoT technologies and morphing via anticipation using relevance vector machines is presented in [39]. Lastly, a method that involves the use of genetic algorithms in smart city partitions—on a scale of neighborhood or region—has been introduced and shown to be effective in enhancing the degree of privacy of the consumption patterns of the partition members.

Overall, the privacy of consumption patterns at the level of hiding individual features in the pattern curvature is an area that has fallen off sight except for a very limited number of works utilizing AI. However, it is expected that the interest in this area will only increase with the penetration of smart city technologies and the presence of third observers that handle the consumption data.

8.4 Discussion

In this chapter, the energy consumption in smart cities is discussed. The increasing adoption of smart city information technologies will transform several of the services as we know them right now into smart components. The cornerstone of smart cities is the continuous energy usage, a term known as smart energy. As it was presented earlier, smart energy encompasses all types of energy sources and not only electricity. It is this interplay of various types of energy that allows the service to be "smart" because energy end-users will be able to select among various types and create their energy consumption portfolio.

However, the development of smart energy services comes with challenges that need to be considered and solutions have to be identified. In this chapter, we identified a set of challenges that arise in smart cities pertained to the deployment of smart energy networks and proposed AI solutions for each challenge have been discussed.

As a general trend, it was identified that the presented challenges impose complex problems that need to be addressed. It was identified that several solutions have been proposed within the context of electrical energy (smart grids) with only limited cases taking into consideration the framework of a smart city. The smart city is a "Dedalus" system comprised of several components with each component composed of several complex subsystems (see Fig. 8.1 where the smart energy component is composed of several complex networks). Thus, there is a big room for using AI tools to manage this highly complex system of a smart city.

In sum, energy consumption is essential in developing the smart cities of the future and will be tackled with smart energy services. However, there are research challenges that need to be further explored and satisfactory solutions will be identified. It is only expected that with the increasing penetration of smart energy those challenges will be getting more attraction from the research community, while the role of AI in smart cities will get a dominant role in energy consumption.

References

1. R.E. Hall, B. Bowerman, J. Braverman, J. Taylor, H. Todosow, U. Von Wimmersperg, *The Vision of a Smart City* (No. BNL-67902; 04042) (Brookhaven National Lab., Upton, NY, US, 2000)
2. I. Yaqoob, I.A.T. Hashem, Y. Mehmood, A. Gani, S. Mokhtar, S. Guizani, Enabling communication technologies for smart cities. IEEE Commun. Mag. **55**(1), 112–120 (2017)
3. J.R. Ferrer, Barcelona's Smart City vision: an opportunity for transformation. Field Actions Sci. Rep. (Special Issue 16), 70–75 (2017)
4. R.P. Dameri, Smart city definition, goals, and performance. In *Smart City Implementation* (Springer, Cham, 2017), pp. 1–22
5. J. Macke, R.M. Casagrande, J.A.R. Sarate, K.A. Silva, Smart city and quality of life: citizens' perception in a Brazilian case study. J. Clean. Prod. **182**, 717–726 (2018)
6. T. Bakıcı, E. Almirall, J. Wareham, A smart city initiative: the case of Barcelona. J. Knowl. Econ. **4**(2), 135–148 (2013)
7. A. Alexandru, M. Ianculescu, I.A. Marinescu, T.D. Popescu, Shaping the digital citizen into a smart citizen on the basis of IoT capabilities, in *2019 22nd International Conference on Control Systems and Computer Science (CSCS)* (IEEE, 2019), pp. 707–714
8. B.N. Silva, M. Khan, K. Han, Towards sustainable smart cities: a review of trends, architectures, components, and open challenges in smart cities. Sustainable Cities Soc. **38**, 697–713 (2018)
9. S.M. Amin, A.M. Giacomoni, Smart grid, safe grid. IEEE Power Energy Mag. **10**(1), 33–40 (2011)
10. E. O'Dwyer, I. Pan, S. Acha, N. Shah, Smart energy systems for sustainable smart cities: current developments, trends and future directions. Appl. Energy **237**, 581–597 (2019)
11. B. Mahdad, K. Srairi, Blackout risk prevention in a smart grid based flexible optimal strategy using Grey Wolf-pattern search algorithms. Energy Convers. Manage. **98**, 411–429 (2015)
12. C. Balakrishna, Enabling technologies for smart city services and applications, in *2012 Sixth International Conference on Next Generation Mobile Applications, Services and Technologies* (IEEE, 2012), pp. 223–227
13. M. Alamaniotis, L.H. Tsoukalas, Implementing smart energy systems: integrating load and price forecasting for single parameter based demand response, in *2016 IEEE PES Innovative Smart Grid Technologies Conference Europe (ISGT-Europe)* (IEEE, 2016), pp. 1–6
14. S. Giest, Big data analytics for mitigating carbon emissions in smart cities: opportunities and challenges. Eur. Plan. Stud. **25**(6), 941–957 (2017)
15. D.S. Chai, J.Z. Wen, J. Nathwani, Simulation of cogeneration within the concept of smart energy networks. Energy Convers. Manage. **75**, 453–465 (2013)
16. L. Kouchachvili, E. Entchev, Power to gas and H2/NG blend in SMART energy networks concept. Renewable Energy **125**, 456–464 (2018)
17. A. Caragliu, C. Del Bo, P. Nijkamp, Smart cities in Europe. J. Urban Technol. **18**(2), 65–82 (2011)
18. T. Nam, T.A. Pardo, Smart city as urban innovation: focusing on management, policy, and context, in *Proceedings of the 5th International Conference on Theory and Practice of Electronic Governance* (2011), pp. 185–194
19. A. Nasiakou, M. Alamaniotis, L.H. Tsoukalas, G. Karagiannis, A three-stage scheme for consumers' partitioning using hierarchical clustering algorithm, in *2017 8th International Conference on Information, Intelligence, Systems & Applications (IISA)* (IEEE, 2017), pp. 1–6
20. M. Alamaniotis, Morphing to the mean approach of anticipated electricity demand in smart city partitions using citizen elasticities, in *2018 IEEE International Smart Cities Conference (ISC2)* (IEEE, 2018), pp. 1–7
21. R. Patrick, N. Bourbakis, Surveillance systems for smart homes: a comparative survey, in *2009 21st IEEE International Conference on Tools with Artificial Intelligence* (IEEE, 2009), pp. 248–252

22. H.J. Visser, V. Pop, J.H.G. Op het Veld, R.J.M. Vullers, Remote RF battery charging, in *PowerMEMS 2010: 10th International Workshop Micro & Nanotechnology for Power Generation and Energy Conversion Applications, Leuven, Belgium, 30 November-3 December 2010* (IET 2011)
23. A. Navidi, F.A.S. Khatami, Energy management and planning in smart cities. CIRED-Open Access Proc. J. **2017**(1), 2723–2725 (2017)
24. M. Alamaniotis, L.H. Tsoukalas, Integration of price anticipation and self-elasticity for hour-ahead electricity bidding and purchasing, in *9th Mediterranean Conference on Power Generation, Transmission, Distribution, and Energy Conversion:* (MEDPOWER 2014) (Piraeus, Greece, 2014), pp. 1–7
25. Z. Allam, Z.A. Dhunny, On big data, artificial intelligence and smart cities. Cities **89**, 80–91 (2019)
26. B. Hrnjica, A.D. Mehr, Energy demand forecasting using deep learning." *Smart cities performability, cognition, & security*. (Springer, Cham, 2020), pp. 71–104
27. M. Alamaniotis, Neuro-kernel-machine network utilizing deep learning and its application in predictive analytics for smart cities, in *Advances in Data Science: Methodologies and Applications*, eds. by G. Phillips-Wren, A. Esposito, L.C. Jain (Springer, Berlin, 2020), Chapter 14, pp. 293–307
28. C.J. Huang, P.H. Kuo, A deep cnn-lstm model for particulate matter (PM2. 5) forecasting in smart cities. Sensors **18**(7), 2220 (2018)
29. M. Alamaniotis, L.H. Tsoukalas, Fuzzy multi-kernel approach in intelligent control of energy consumption in smart cities, in *29th IEEE International Conference on Tools with Artificial Intelligence* (ICTAI 2017) (Boston, MA, USA, 2017), pp. 1021–1028
30. S. Karnouskos, Demand side management via prosumer interactions in a smart city energy marketplace, in *2011 2nd IEEE PES International Conference and Exhibition on Innovative Smart Grid Technologies* (IEEE, 2011), (pp. 1–7)
31. N. Bourbakis, L.H. Tsoukalas, M. Alamaniotis, R. Gao, K. Kerkman, Demos: a distributed model based on autonomous, intelligent agents with monitoring and anticipatory responses for energy management in smart cities. Int. J. Monit. Surveillance Technol. Res. **2**(4), 81–99 (2014)
32. P. Dimitroulis, M. Alamaniotis, Residential energy management system utilizing fuzzy based decision-making, in *IEEE Texas Power and Energy Conference* (College Station, TX, USA, 2020), pp. 6
33. C. Lazaroiu, M. Roscia, D. Zaninelli, Fuzzy logic to improve prosumer experience into a smart city, in *2018 International Conference on Smart Grid (icSmartGrid)* (IEEE, 2018) pp. 52–57
34. M.Alamaniotis, L.H. Tsoukalas, N. Bourbakis, Virtual cost approach: electricity consumption scheduling for smart grids/cities in price-directed electricity markets, in *IISA 2014, The 5th International Conference on Information, Intelligence, Systems and Applications* (IEEE, 2014), pp. 38–43
35. N. Petrović, Đ Kocić, Data-driven framework for energy-efficient smart cities. Serb. J. Electr. Eng. **17**(1), 41–63 (2020)
36. M. Alamaniotis, N. Bourbakis, L.H. Tsoukalas, Enhancing privacy in smart cities through morphing of anticipated demand utilizing self-elasticity and genetic algorithms, in *Sustainable Cities and Society*, vol. 46 (Elsevier, 2019) pp. (101426)1–12
37. M. Alamaniotis, N. Bourbakis, L.H. Tsoukalas, Anticipatory driven nodal electricity load morphing in smart cities enhancing consumption privacy, in *IEEE PES PowerTech Conference* (Manchester, UK, 2017), pp. 1–6
38. M. Alamaniotis, L.H. Tsoukalas, M. Buckner, Privacy-driven electricity group demand response in smart cities using particle swarm optimization, in *IEEE International Conference on Tools with Artificial Intelligence* (ICTAI 2016) (San Jose, CA, USA, 2016), pp. 946–953
39. F. Belligianni, M. Alamaniotis, A. Fevgas, Tsompanopoulou, P. Bozanis, L.H. Tsoukalas, An internet of things architecture for preserving privacy of energy consumption, in *10th Mediterranean Conference on Power Generation, Transmission, Distribution and Energy Conversion* (Med Power 2016), Belgrade, Serbia, November 6–9, 2016.

Chapter 9
How to Make Different Thinking Profiles Visible Through Technology: The Potential for Log File Analysis and Learning Analytics

Gyöngyvér Molnár

Abstract The aims of this paper are to show how innovations in technology-based assessment and the options available to us in educational data mining and log file analysis can provide answers to research questions which remain unanswered with traditional assessment techniques and can quantify qualitative differences by making thinking visible. The paper outlines how log file analysis and learning analytics have entered the mainstream in the last ten years, and it shows how log file analysis can be used to make different thinking behaviour and strategy usage among students visible in interactive problem-solving environments within the MicroDYN framework. The sample for the experimental study was drawn from first-year university students (N = 1682, M_age = 19.9, SD_age = 2.52). All the interactive problems were administered online via the eDia platform. Time-on-task and latent class analysis was employed to find students whose thinking strategies show similar patterns and who have thus proved to be qualitatively similar thinkers during the problem-solving process. Results confirmed that educational data mining and log file analysis can be used to make students' thinking visible and to quantify qualitative differences in thinking. This study contributes to an understanding of how students interact with unknown and complex problems and how they think while solving them.

9.1 Introduction

Technology offers new opportunities in educational assessment and research. Paper-and-pencil and face-to-face assessment and training methods provide the researcher with very few quantitative and qualitative indicators, such as test scores and subjective feedback from students [13]. Technology offers not only the option of measuring and fostering constructs that would be impossible to measure and enhance through traditional techniques, but also the opportunity to analyse data logged and collected

G. Molnár (✉)
Department of Learning and Instruction and MTA–SZTE Research Group on the Development of Competencies, University of Szeged, Szeged, Hungary
e-mail: gymolnar@edpsy.u-szeged.hu

M. Virvou et al. (eds.), *Advances in Artificial Intelligence-based Technologies*,
Learning and Analytics in Intelligent Systems 22,
https://doi.org/10.1007/978-3-030-80571-5_9

during the testing/learning session (e.g. time-on-task, number of attempts within the same task and number of clicks,see [11, 19, 24] and quantify qualitative differences [15, 16] for a deeper and better understanding of the fine mechanisms of the phenomenon under examination.

These new opportunities have resulted in changes in previous learning theories and practices and the development of qualitatively new theories and teaching methods [1]. The new technologies provide options in logging students' actions during the learning process that have not been available before [12].

Beyond opportunities to take a closer look at direct output data and the structure of the construct under examination, log file analysis and learning analytics enable us to predict which types of activities would be the most beneficial for different students, thus contributing significantly to the personalization of education [23]. They make it possible not only to confirm that a given learning unit has been mastered, but also to monitor the learning activity in real time. Based on these data, both computer-controlled and human-driven techniques can be used to better tailor education to the needs of learners, moving away from the one-size-fits-all approach [23]. These innovations are expected to revolutionize education and suggest that learning analytics may become one of the most significant developments in the twenty-first century [9].

The idea of using assessment-based data to make thinking visible is rooted in the following idea introduced by John [8]: use assessment and feedback to make learning visible. The present paper introduces an exploratory study with the aim of examining students' learning behaviour and thinking procedure in a dynamically changing problem-solving environment using log file analysis. It provides both a qualitative and quantitative description of the thinking procedures students use in mapping dynamically changing, simulated problems.

The way log file analysis and learning analytics have entered the mainstream in the past decade is summarized, thus opening new possibilities and raising new research questions about students' way of learning and thinking, for example, how log file analysis and learning analytics can be used to provide more elaborated and qualitatively different feedback on students' cognitive development beyond the simple test score and how all this can contribute to a significantly better understanding of the different ways students think while learning in a dynamically changing problem-solving environment.

9.2 The Development of Log File Analysis and Learning Analytics

As a result of major developments in large-scale international assessments (e.g. OECD PISA, IEA TIMSS and IEA PIRLS), educational assessment has been among the most dynamically developing areas in education in the last two decades. Technology-based assessment and enhancement have supplanted traditional assessment [15] and training methods. They provide automated feedback, new item formats, real-life situations and authentic tasks. Technology enables us (1) to measure 21st-century skills using interactions, dynamism, virtual worlds and collaboration [17, 18] and (2) to gather information via log files beyond the answer data (e.g. response time, time-on-task, repetition and eye movements).

Before the emergence of learning analytics, empirical research mainly concentrated on learning outcomes with considerable limitations in terms of data and sample size. Learning analytics operates with massive, originally unstructured databases and facilitates the discovery of basically novel, potentially useful and certainly interesting information from large amounts of logged data. This paper will demonstrate just some of the milestones in this developmental process.

The possibilities of log file analysis have inspired new research endeavours and resulted in new terminology, such as educational data mining, learning analytics and big data. While 2010 saw only eight publications on educational data mining in education and psychology on Scopus, this number significantly increased by 2019, with 15.5 times more articles tied to educational data mining that year. No papers were published on learning analytics in the social sciences or psychology in 2010, while that number grew to 410 in 2019. The largest development happened in big data. The number of publications has risen immensely in the last ten years from six (in 2010) to 1647 (in 2019). Please note that these figures only refer to papers published in the social sciences and psychology.

In 2011, the first international conference on learning analytics and knowledge was launched (https://tekri.athabascau.ca/analytics/) with the aim of focusing on the integration of the technical and social/pedagogical dimensions of learning analytics. In the same year, Nikolaos G. Bourbakis launched the International Journal on Artificial Intelligence Tools.

2012 witnessed the establishment of the 27th special interest group within the biggest European association in education (the European Association on Learning and Instruction, EARLI), the Online Measures of Learning Processes SIG, with the aim of (1) gathering people who look at learning analytics from different perspectives, either research or practice, but also (2) supporting developments in this area.

In 2013, the Journal of Learning Sciences devoted a special issue to learning analytics and computational techniques for detecting and evaluating patterns in learning [12].

2014 saw the launch of the Journal of Learning Analytics. This basically interdisciplinary journal aims to connect researchers, developers and practitioners who share a common interest in learning analytics using data traces to better understand and improve learning through the creation and implementation of new tools and techniques.

In 2015, the Joint Research Centre initiated a study with the support of the European Commission to find evidence for the use of learning analytics in education and to understand the state of the art in the implementation of learning analytics for education and training in both formal and informal settings [22].

In 2016, Avella et al. published a systematic literature review in learning analytics. They included 44 articles in the analysis with the aim of providing an overview of the methods, benefits and challenges of using learning analytics in higher education. They observed that researches on learning analytics use a variety of methods, including visual data analysis techniques, social network analysis, and semantic and educational data mining, including prediction, clustering, relationship mining, discovery with models and separation of data for human judgement. These methods are employed to analyse data and overcome numerous challenges, such as data tracking, collection, evaluation and analysis.

2018 marked the publication of a special issue of the Journal of Interactive Learning Environments on educational big data and learning analytics because, according to the editors, educational big data and learning analytics had entered the mainstream in numerous research fields. In the same year, [20] analysed 252 papers on learning analytics in higher education published between 2012 and 2018. They focused on four propositions validated earlier: whether learning analytics improves learning outcomes, supports learning and teaching, is deployed widely and, finally, is used ethically. The result of the meta-analysis indicated that despite the fact that it offers great potential for improving learner practice, not much transfer to practice can currently be seen and there is little evidence of the four propositions, whether it is learning outcomes (9%) or improvements in learning support and teaching (35%) or the third (6%) or fourth (18%) proposition.

A year later, in 2019, the EARLI Conference of Assessment and Evaluation SIG concentrated on learning analytics, and another systematic meta-review [5] was published, including a review of 901 learning analytics papers. According to their analysis, the majority of the publications were focused on proposing concepts and frameworks and not on conducting actual data analysis. Generally, collecting small datasets in K–12 predominates. [2] reviewed 402 studies on educational data mining and learning analytics. They found that specific educational data mining and learning analytics techniques could offer the best means of solving certain learning problems. The application of these opportunities in higher education can aid us in developing a student-focused strategy.

Based on the large quantity of research in log file analysis, it can be concluded that there has been a shift towards a deeper understanding of students' learning experiences for the last ten years. The present paper contributes to this issue and provides data-driven examples on how log file analysis and learning analytics can be used to better education, to learn more beyond the traditional data and to understand the differences in students' thinking while acquiring new information in a dynamically changing problem-solving environment.

9.3 Analysing Log File Data in Researching Dynamic Problem-Solving

Problem-solving skills are among the most extensively studied transversal skills over the last decade. Several theoretical models and definitions have been published with common aspects: a problem is characterised by a gap between the current and the goal state with no immediate solution available (Mayer and Wittrock 1996). Nowadays, taking advantage of the possibilities of computer-based assessment, the research on problem-solving disregards content knowledge and domain-specific processes [14], moves from static to dynamic problem-solving and makes use of the possibilities of log file analysis. Research endeavours to examine performance only. The final score (i.e. correct or incorrect) is not sufficient because it cannot establish principles on how learners interact with the problem environment using the affordances of technology to solve a problem [3]. Log files can provide direct insights into how results are produced, how a student interacts with the problem and how this results in specific mastery or non-mastery of a task [7].

From an empirical perspective, studies addressing the field of assessment of problem-solving five years ago largely focused on time-on-task as the central behavioural indicator (Goldhammer et al. 2014; Greiff et al. 2016). Goldhammer et al. (2014) showed that more time spent on a problem in a problem-solving environment was associated with higher performance; however, the relationship between time spent on a problem and actual performance did not seem to be linear (Greiff et al. 2016). According to Greiff et al. (2016), this relationship seems to be quadric. Too long or too short times were both associated with poor performance.

Nowadays, there have been more studies published with regard to problem-solving and log file analysis, which have applied other process indicators of test-taking behaviour, such as specific exploration strategies. Scherer and Tiemann (2012) showed that pre-existing strategic knowledge is associated with the exploration strategy that is applied in a problem-solving environment, which in turn is related to overall problem-solving performance. As regards the different kinds of exploration strategy usage, a specific strategy that has often been investigated in problem-solving research, especially for problems developed in the MicroDyn approach, is the application of the VOTAT strategy (vary-one-thing-at-a-time). When applying the VOTAT strategy, only one variable is manipulated, while all the other variables are non-manipulated to determine the effect of the varied variable on dependent outcomes. Wüstenberg et al. (2012), [14] and others showed that application of the VOTAT strategy is strongly related to problem-solving performance in the MicroDyn approach. Moreover, [14] identified different types of VOTAT strategies and empirically confirmed that not all VOTAT strategies result in high problem-solving performance. The isolated variation strategy—when students examined the effect of input variables on output variables independently, that is, when only one input variable is varied, while all other variables are kept constant (the principle of isolated variation; [21]—was the most effective strategy and comprised 45% of the VOTAT strategies employed. It resulted in the correct solution to the problem with higher probability, independently of problem complexity or the grade the students were in.

To sum up, log file analysis makes it possible to cluster the students' problem-solving behaviour and exploration strategy usage at the beginning of the problem-solving process. Log file analysis combined with latent class analysis can make students' different thinking profiles visible; that is, the current article wishes to contribute insights into students' approaches and profiles to explore and solve dynamic problems related to minimal complex systems. The research investigates learners' interactions with a dynamic problem-solving environment, as well as defines learners' learning and thinking profiles during problem exploration before deciding on the necessity of training and the kind of training that may be helpful to enhance those who need assistance. Based on latent profiles, it tests time-on-task separately to check if any of the relations identified earlier are invariant across the different profiles.

9.4 Extracting, Structuring and Analysing Log File Data to Make Different Thinking Profiles Visible

9.4.1 Aims

This study uses log file analysis to empirically test different behaviour and strategy usage among first-year university students and make the differences in their thinking visible while they solve dynamic problems developed with the MicroDyn approach. In the first section of the empirical part of the paper, the paper aims: (1) to introduce how the originally unstructured log files collected during the problem-solving process became structured data; (2) to test the effect of the number of manipulations made, time-on-task, type of thinking and type of strategy applied in the exploration phase on actual problem-solving achievement; (3) to outline the possibilities in latent class analysis on structured log data to find similar patterns to make students' thinking visible; and (4) to separately test the effect of the number of manipulations made and of time-on-task based on the latent profiles to check if any of the relations identified are invariant across the different profiles.

9.4.2 Methods

9.5 Participants

The sample for the study was drawn from first-year university students (N = 1682). The students were 19.9 years old on average (SD_age = 2.5). The proportion of men and women was about 49 to 51 per cent. 43.4% of the target population participated in the assessment. Participation was voluntary; as an incentive, students received credit for successful completion of the test.

9.6 Instruments

The MicroDYN approach was employed to develop a third-generation online measurement device for dynamic problem-solving. All the problems have different fictitious cover stories to avoid the use of existing factual knowledge during the problem-solving process. That is, the content of the problems is not based on real causal relationships or on real knowledge; it is completely artificial. Content knowledge is therefore no help to the students in solving the problems successfully.

To solve these problems, the problem-solver must engage in and interact with the system by freely manipulating the values of the input variables (Greiff and Funke 2010) and determining the effect of this on the system and on the level of the output variables with feedback provided by the computer (Fischer et al. 2012). The interactions between the problem situation and the problem-solver can only be observed in a computerized environment based on log data collected during test administration (see Fig. 9.1). During this first phase, the problem-solver is asked to depict the new knowledge acquired during the exploration phase by indicating the connections identified between the different variables on a concept map onscreen, that is, by expressing the mental model structure on an external symbolic representation of the system.

In the second phase of the problem-solving process, students are asked to use the system, changing the values of the input variables and thus reaching the given target area in no more than four steps and within 180 s. In this case, the right concept map is provided onscreen.

The complexity of the problems, that is, the number of input and output variables and number of relations between them, increased on the test. The last problems on the test also contain internal dynamics (eigendynamics), that is, autoregressive dependencies. In other words, the problem environments change without any intervention or any manipulation of the input variables (see Table 9.1, first column).

The test prepared for this study consists of 24 items with varying difficulty levels. For example, in the first phase, the knowledge acquisition phase, of an easy problem, students are asked to observe how changing the values of two different kinds of drink mix impacts the value of the sweetness of the drink (see Fig. 9.1), and, in the second phase, they are asked to reach the required sweetness value by setting the proper values of the two kinds of drink mix. In a more difficult problem, students are asked to comprehend more complex relationships between three different training methods used by basketball players as well as motivation, power of the throw and exhaustion. (For more examples of similar dynamic problem-solving items, see Greiff et al. 2013; OECD 2014.)

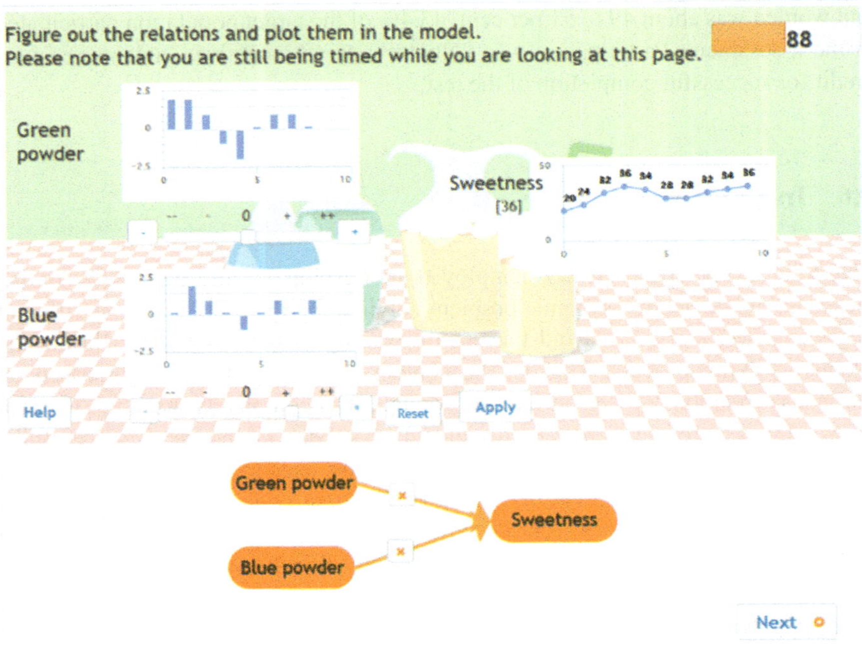

Fig. 9.1 Screenshot of the MicroDYN task Drink Mix. Exploration of the problem – the first phase of the problem-solving process. The controllers of the input variables range from "- -" (value = -2) to " + + " (value = + 2). The concept map is shown onscreen

Table 9.1 The relationship between number of manipulations made, time-on-task and achievement scores based on the problems administered on the test

Number of the problem (complexity)	r (achievement/ number of manipulations)	r (achievement/ time-on-task)
1. (2–1–2–0)	−0.098**	−0.203**
2. (2–2–2–0)	n.s	−0.085**
3. (2–2–2–0)	0.082**	−0.068**
4. (3–2–3–0)	n.s	−0.159**
5. (3–2–3–0)	0.136*	−0.070**
6. (3–3–3–0)	n.s	−0.142**
7. (3–3–4–0)	n.s	−0.085**
8. (3–3–4–0)	n.s	−0.135**
9. (3–2–1–1)	0.237**	0.161**
10. (3–2–2–1)	0.290**	0.106**
11. (3–3–3–1)	0.310**	0.174**
12. (3–3–3–1)	0.402**	0.261**

Note Complexity: number of input and output variables, number of relations and the presence of eigendynamics. Level of significance: *$p < 0.05$, **$p < 0.01$

9.7 Procedures

The dynamic problems were administered online via the eDia system. At the beginning, students were provided with instructions on how to use the system, including a warm-up task. At this point, they had the opportunity to learn how to use the test environment: (1) a yellow bar indicates how far along they are on the test; (2) they click on the Apply button to test a setting for the input variables; (3) they click on the Next button to move on to the next phase of the problem and, after the second phase, on to the following problem; (4) they first click on the starting point of the arrow, then on the end point of it to draw arrows on the concept map; they click on an "x" on the arrow to delete the arrows; and (5) they receive immediate feedback on their performance after solving the last problem.

The assessment took place in the ICT lab at the University of Szeged Learning and Information Centre. It took approximately 45 min to complete. Students' activity was logged, coded and then clustered for comparison. A latent profile analysis was conducted to ascertain exploration strategies and different thinking routes employed by our latent classes as they worked to understand the problems.

In the first phase of the problem-solving process, students had the opportunity to manipulate the system and apply as many exploration steps as they wished. In the background, each step was stored in the log data. In order to structure the log data, the continuum of the problem-solving process was divided into scoreable phases with a labelling procedure. A three-level labelling procedure was defined for the most effective clustering, distinguishing students' use of three different kinds of input behaviour within a single problem: full input behaviour (with every activity and manipulation with the sliders and the application button included), basic input behaviour (which consists of manipulations where problem-solvers were able to acquire new information on the system as well as timing, that is, the order of the trials executed) and minimal input behaviour (a simple list of activities that were out of order, with which problem-solvers were able to acquire new knowledge from the system). The unit of this labelling process was a setting of the input variables (a trial), which was tested at the same time by clicking on the application button during the first phase of a problem. The sum of these trials within the same problem is called the input behaviour, which was called a strategy if it exhibited meaningful regularities.

Each of the manipulations and each of the trials were labelled according to [13] labelling system (please note that the numerical labels are neither scores nor ordinal or metric information), which is based on the number of changed variables, the familiarity level of the new settings and the amount of new information:

- The trial was labelled + 1 if only one variable was changed and its relationship to the output variables was unknown, that is, if its effect cannot have been identified from previous trials, while the other variables were set at a neutral value like zero.
- The trial was labelled + 2 if only one variable was manipulated and its relationship to the output variables was also unknown. The values of the other input variables were not set at zero, but at a setting used earlier.

- The trial was labelled + 3 if the requirements of the + 2 case were fulfilled and the combination had not been attempted earlier.
- The trial was labelled + A if everything was maintained in a neutral (zero) position.
- The trial was labelled –X if more than one input variable was manipulated at the same time and its relationship to the output variables was unknown. This trial resulted in no additional information on the problem.
- The trial was labelled –0 if the same setting, that is, the same manipulation, had already been employed earlier within the problem. This trial also offered no additional information on the problem.
- The trial was labelled + 0 if a new setting was employed, but the extracted information was known from previous settings. This trial also resulted in no additional information on the system.
- The trial was labelled + 4 if several variables were changed, but it was theoretically possible to count the effect of the input variables based on information from previous and present settings.
- An extra code (+5) was used for the basic model when the problem-solver was able to figure out the structure of the scenario based on information learned in the last trial.

After the labelling procedure, these new labels garnered scores of 0, 1 or 2, according to the application level of the principle of isolated variation strategy, which is the most efficient exploration strategy in the case of minimal complex systems, like the MicroDyn problems. By using the principle of isolated variation strategy, the problem-solver systematically varies only one input variable at a time, while the others remain unchanged. The most obvious isolated variation strategy is when only one input variable is manipulated systematically and as a result differently from the neutral level and all the other input variables are maintained at the neutral level.

In the scoring procedure, 0 points were given if no isolated variation was applied at all, 1 point for partial isolated variation (isolated variation strategy was applied for some but not all of the input variables), and 2 points for full isolated variation (this was applied to all the input variables).

The actual achievement of the student, that is, the structure of the concept map (in the first phase) and successful application of the system (in the second phase), was scored dichotomously. One point was given for a completely correct model on the concept map; otherwise, the response was assigned a score of 0. In the second phase, students received 1 point if they managed to reach the given target values of the output variables within 180 s and at most in four steps; otherwise, their response was assigned 0 points.

9.8 Results

The internal consistency of the test was high based on the traditional scoring methodology (Cronbach's $\alpha = 0.89$). However, the reliability of the test was higher based on the recoded log data and the exploration behaviour of the students (Cronbach's $\alpha = 0.94$). The reliability did not increase when both the traditional scoring and the log data-based information were used (Cronbach's $\alpha = 0.94$). That is, the test results in the case of the dynamic problem-solving test with the MicroDyn approach are more reliable with a focus on the behaviour of the students via the labelled log data rather than using the traditional item scores in the analysis exclusively.

Based on the manifest achievement data, there was a low, but significant correlation between the number of manipulations made and students' achievement in the first phase of the problems ($r = 0.276$, $p < 0.001$) on the sample level; that is, students who tended to attempt more cases and played more with the problem situation to identify the relationships between input and output variables and the behaviour of the problem achieved significantly higher and were more successful at structural model building. This correlation changed according to the complexity of the problems and the number of problems explored earlier in the test (see Table 9.1). There was almost no relation observed in problems containing only direct effects and placed at the beginning of the test. The correlation increased significantly in problems with eigendynamics.

The time spent exploring the problem situation did not correlate significantly with the students' actual achievement in the first phase of the problems ($r = 0.045$, $p < 0.05$). Thus, without a deeper analysis of the log data, students' performance and problem-solving efficiency were clearly less determined by time spent recognising the problem on the sample level. However, after a closer look at the log data and analysing correlations on the level of the problems, the correlation was certainly weak, but significant and negative for problems with direct effects. It was only positive for the most complex problems (see Table 9.1).

In the next step, the type and quality of the manipulations made and the type of thinking procedures were also involved in the analysis (see Table 9.2). The correlation between the number of manipulations made and the quality of the regularity in the manipulations was tested to ascertain whether this manipulation covers a theoretically grounded strategy for exploring the problem environment or not. The relation between the number of manipulations and the application of a theoretically grounded exploration strategy increased during the test. It requires further study to ascertain whether this was caused by the learning effect during the test-taking procedure or by the complexity of the problems. A similar pattern was observed between time-on-task and the application of a theoretically grounded strategy. The correlation coefficient was significant throughout the test, but it grew stronger and stronger with time and complexity.

The relation between the application of a theoretically good thinking strategy and high performance also changed with the complexity of the problem and the learning effect during the testing procedure ($r = 0.522$, $p < 0.001$ on test level). First, it can

Table 9.2 The relationship between number of manipulations made, time-on-task and achievement scores based on the problems administered on the test

Number of the problem (complexity)	r (theoretically grounded strategy/ number of manipulations)	r (theoretically grounded strategy/ time-on-task)	r (theoretically grounded strategy/ achievement)
1. (2–1–2–0)	0.250**	0.138**	n.s
2. (2–2–2–0)	0.227**	0.149**	0.231**
3. (2–2–2–0)	0.235**	0.146**	0.278**
4. (2–2–2–0)	0.247**	0.108**	0.405**
5. (3–2–3–0)	0.302**	0.182**	0.355**
6. (3–3–3–0)	0.308**	0.118**	0.417**
7. (3–3–4–0)	0.315**	0.184**	0.460**
8. (3–3–4–0)	0.263**	0.142**	0.478**
9. (3–2–1–1)	0.308**	0.247**	0.139**
10. (3–2–2–1)	0.336**	0.262**	0.157**
11. (3–3–3–1)	0.374**	0.320**	0.199**
12. (3–3–3–1)	0.375**	0.308**	0.203**

Note Complexity: number of input and output variables, number of relations and the presence of eigendynamics. Level of significance: *$p < 0.05$, **$p < 0.01$

be concluded that a theoretically grounded exploration strategy did not always result in high achievement. Second, the relation became stronger for both problem types (only direct effect vs. eigendynamics) as a result of a learning effect; however, it dropped immensely on problems with eigendynamics.

Generally, based on the log file analysis, it can be concluded that students who put more effort into exploring the problem environment achieved higher than others. Students were able to learn during the testing session, and their thinking procedure, the regularity of it and their awareness changed even after solving a few basically similarly structured problems.

The type and effectiveness of the exploration and thinking strategies in a dynamic problem-solving environment were analysed by means of latent profile analysis using the labelled and recoded log file data collected during the first phase of the problem-solving process. Problems with only a direct effect (the first eight problems) or internal dynamics (the last four problems) have been separated.

Different latent class profiles were hypothesised and examined on the basis of students' manipulation and thinking procedures, searching for patterns indicating the usage of the principle of isolated variation to some extent while exploring the dynamic problem scenarios.

The information theory criteria used (AIC, BIC and aBIC) indicated a continuous decrease in a growing number of latent classes. The likelihood ratio statistical test (Lo–Mendell–Rubin Adjusted Likelihood Ratio Test) showed the best model fit for the 4-class model. The entropy-based criterion reached the maximum values for the

Table 9.3 Information theory, likelihood ratio and entropy-based fit indices for latent class analysis based on problems with a direct effect

Number of latent classes	AIC	BIC	aBIC	Entropy	L–M–R test	p
2	7751	7930	7825	0.973	3544	0.0001
3	7389	7660	7501	0.926	393	0.0001
4	7257	7620	7408	0.951	164	0.0008
5	7241	7696	7429	0.944	50	1.00

Note AIC = Akaike Information Criterion; BIC = Bayesian Information Criterion; aBIC = adjusted Bayesian Information Criterion. L–M–R test = Lo–Mendell–Rubin Adjusted Likelihood Ratio Test. The best fitting model solution is in italics

2-class solutions, but it was also significant for the 3- and 4-class solutions. The entropy index for the 4-class model showed that 95% of the first-year students were accurately categorized based on their class membership (Table 9.3).

Based on the findings (AIC = 7257, BIC = 7620, aBIC = 7408, Entropy = 0.95, LMR test = 164, $p < 0.001$), a distinction can be made between four qualitatively different groups by exploring dynamic problems with direct effects (see Fig. 9.2): (1) proficient strategy users (80.2%); (2) almost high performers on the easiest problems, but low performers on the complex ones with a slow learning effect (4.7%); (3) rapid learners (7.8%), the most valuable group from an educational perspective; and, finally, (4) low to intermediate performers on the easiest problems, but non-performers on the complex ones with a very slow learning effect (7.3%). No so-called intermediate strategy users were found, a result which is consistent with [13] findings on students in Grades 3–12.

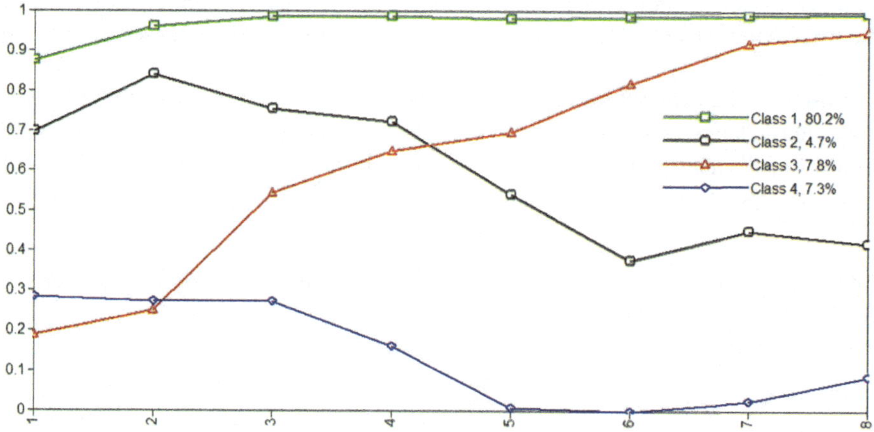

Fig. 9.2 Latent class profiles based on the log data collected in the exploration phase of the problem-solving process for: proficient explorers (Class 1); almost high performers on the easiest problems, but low performers on the complex ones with a slow learning effect (Class 2); rapid learners (Class 3); and low to intermediate performers on the easiest problems, but non-performers on the complex ones with a very slow learning effect (Class 4)

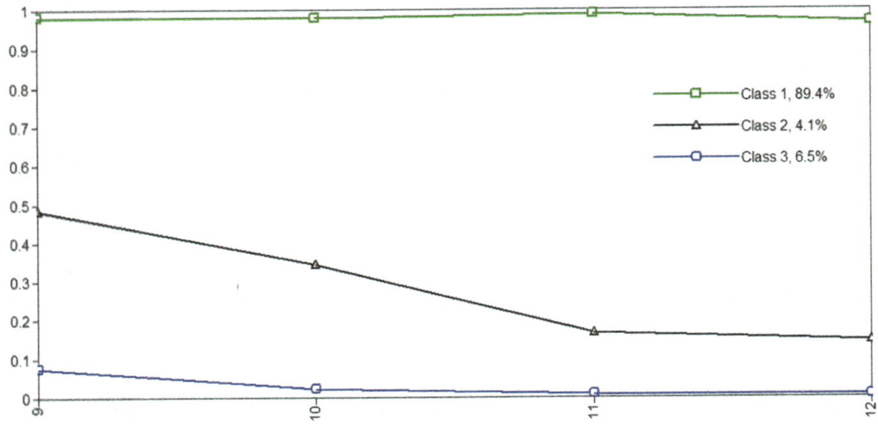

Fig. 9.3 Latent class profiles based on log data collected in the exploration phase of the problem-solving process for (1) proficient explorers, (2) intermediate performers on the easiest problems, but low performers on complex ones, and (3) non-performers

The result was significantly different when only focusing on log data collected during the exploration phase of the problems with eigendynamics. Based on the latent profile analysis, only three significantly different thinking patterns were identified (see Fig. 9.3).

Almost 90% of the students applied the isolated variation strategy (beyond the isolated variation strategy to map the exact behaviour of these kinds of problems, students also need to apply the trial labelled + A), while the remainder of the students achieved intermediate or low on the easiest problems and very low on the more complex ones. According to the entropy index for the 3-class model, almost 97% of the first-year students were accurately categorized based on their class membership (Table 9.4).

Based on the patterns in the recoded, labelled and scored log data and information criteria, latent classes were built on students' level of exploration expertise with regard to fully and partially isolated variation during the entire test (twelve problems), independent of the type of problem visible (Fig. 9.4). As expected, it was

Table 9.4 Information theory, likelihood ratio and entropy-based fit indices for latent class analysis based on problems with eigendynamics

Number of latent classes	AIC	BIC	aBIC	Entropy	L–M–R test	p
2	3471	3564	3510	0.991	2050	0.0001
3	3420	3561	3478	0.966	68	0.0106
4	3417	3606	3495	0.946	20	0.1893

Note AIC = Akaike Information Criterion; BIC = Bayesian Information Criterion; aBIC = adjusted Bayesian Information Criterion. L–M–R test = Lo–Mendell–Rubin Adjusted Likelihood Ratio Test. The best fitting model solution is in italics

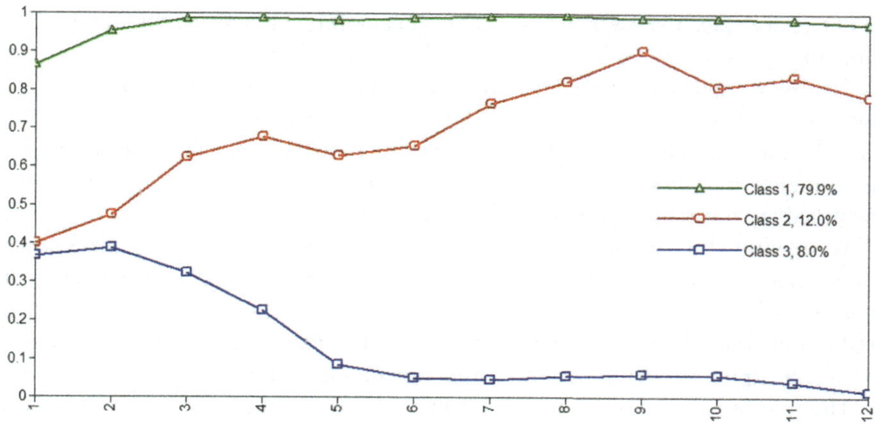

Fig. 9.4 Latent class profiles based on log data collected throughout the test for (1) proficient explorers, (2) rapid learners, and (3) intermediate performers on the easiest problems, but non-performers on the complex ones with a very slow learning effect

possible to identify proficient explorers (almost 80% of the students), who recognised almost from the very first problem that the most effective thinking strategy to explore these kinds of systems is the isolated variation strategy. Based on the log files, their manipulations were almost perfect in grasping the system. Another question is why some of them failed and made the wrong decision about the problem (Mean = 68.1%, SD = 22.6, Mean_time-on-task_exploration = 649, Sum_manipulation = 68, Mean_quality_of_the_manipulation = 99.2%).

It was also possible to identify rapid learners (12% of the students), who form the most interesting group from the perspective of teaching (Mean = 33.2%, SD = 19.6, Mean_time-on-task_exploration = 696, Sum_manipulation = 60, Mean_quality_of_the_manipulation = 86.7%). Finally (see Table 9.5), it was possible to identify students whose exploration strategy was moderately successful on the easiest problems but who did not know what to do on the more complex ones (see Fig. 9.4). This 8% of the students require further help to avoid dropping out

Table 9.5 Information theory, likelihood ratio and entropy-based fit indices for latent class analysis based on problems with eigendynamics

Number of latent classes	AIC	BIC	aBIC	Entropy	L–M–R test	p
2	12,195	12,461	12,305	0.984	5556	0.0001
3	11,461	11,863	11,627	0.930	779	0.0001
4	11,287	11,824	11,509	0.945	223	0.6416

Note AIC = Akaike Information Criterion; BIC = Bayesian Information Criterion; aBIC = adjusted Bayesian Information Criterion. L–M–R test = Lo–Mendell–Rubin Adjusted Likelihood Ratio Test. The best fitting model solution is in italics of university (Mean = 18.9%, SD = 14.6, Mean_time-on-task_exploration = 257, Sum_manipulation = 28, Mean_quality_of_the_manipulation = 32.2%).

Finally, we separately tested the effect of the number of manipulations made and time-on-task based on the latent profiles [(1) proficient explorers, (2) rapid learners and (3) intermediate performers on the easiest problems, but non-performers on the complex ones with a very slow learning effect] to check if any of the relations identified are invariant across the different profiles.

We identified different patterns in each of the groups. On the test and sample levels, there was no correlation between time-on-task and student achievement, but time-on-task played a significant role among the low achievers ($r = 0.228$, $p < 0.01$). However, this correlation proved non-significant among the rapid learners and proficient explorers, resulting in no correlation on the sample level.

The role of number of clicks also changed from group to group. The strongest correlation between the number of manipulations and test-level achievement was observed among the students with a low performer profile ($r = 0.313$, $p < 0.01$). The level of correlation decreased but remained significant with the increasing level of problem-solving proficiency ($r = 0.187$, $p < 0.01$ among the rapid learners; $r = 0.134$, $p < 0.01$ among the high performers).

The quality of the manipulations made also played a much larger role in the low performers' group ($r = 0.506$, $p < 0.01$) than among the rapid learners ($r = 0.347$, $p < 0.01$) or proficient explorers ($r = 0.228$, $p < 0.01$). The change in time-on-task and the number of clicks was not linearly dependent on the expertise level of the students. The strongest correlation was observed among the low performers ($r = 0.739$, $p < 0.01$) and high performers ($r = 0.659$, $p < 0.01$), while there was a lower correlation among the rapid learners ($r = 0.583$, $p < 0.01$). The correlations between the manipulations made and time-on-task or number of clicks differed the most. The correlation proved to be the strongest among the low performers ($r = 0.631$, $p < 0.01$; $r = 0.794$, $p < 0.01$), medium to weak among the rapid learners ($r = 0.316$, $p < 0.01$; $r = 0.506$, $p < 0.01$) and weak among the high performers ($r = 0.062$, $p < 0.05$; $r = 0.117$, $p < 0.01$).

9.9 Discussion

The aim of this study was: to investigate the effect of the number of manipulations made, time-on-task, type of thinking and type of strategy applied in the exploration phase on actual problem-solving achievement; to find patterns in students' problem-solving behaviour to make students' thinking profile visible; and to test the role of these four indicators and invariance across the different thinking profiles.

The results show that the answer to the first question depends on the task: it seems to be disadvantageous to take a long time on the easiest, simplest tasks, while a relatively long exploration time is beneficial on the most complex tasks, taking the final problem-solving achievement into account. The correlation between time-on-task and achievement grew stronger among more complex problems. This result is partly in line with findings by Goldhammer et al. (2014). According to Goldhammer et al. (2014), time spent on a task depends on the type and nature of the task and the cognitive processes required to complete that task. On tasks requiring non-routine behaviour, time-on-task can be interpreted as an indicator of effort, while on tasks requiring only routine behaviour, time-on-task can be explained as an indicator of having trouble solving the problem. In line with this theory, they showed that more time spent on problem-solving tasks was associated with higher performance and time-on-task strongly depended on task difficulty. Naumann et al. (2014) contended that a large amount of time spent on easy problems requiring not much exploration could indicate disorientation or mental overload, resulting in low performance. In contrast, Greiff et al. (2016) argued that the relationship between time spent on a task and problem-solving performance is not linear, but quadric, where times that were too long or too short were associated with poor performance.

If the focus was more on the quality of the exploration and not on the final problem-solving achievement, a longer time resulted in higher-quality manipulation. That is, it can be assumed that the effects of the four process indicators on performance strongly depend on task complexity and difficulty and may change among students characterised by a different thinking profile.

In the next research question based on structured log data, we investigated whether different thinking profiles can be identified. Using latent class analysis, we found three qualitatively different class profiles. The first was that of proficient strategy users, who consistently employed optimal exploration strategies from the very first problem (in 99.2% of the cases) and were also the most successful in interpreting the information extracted from the problem environment (Mean = 68.1%, SD = 22.6). Their achievement is in line with the highest number of manipulations extracted and with a relatively long time spent on task. The question of why there are such big differences in the quality of manipulations made and the final problem-solving achievement calls for further study.

It was also possible to identify rapid learners. These students started out as low to medium performers in their exploration behaviour on the first problems on the test, showed a rapid learning curve afterwards and began to use high-quality exploration strategies. By the end of the test, they reached the same high level of exploration behaviour as the members of the proficient explorers' group, resulting in a relatively high quality of manipulation (86.7%); however, their mean test-level achievement was significantly lower (Mean = 33.2%, SD = 19.6), much lower than expected based on their exploration behaviour. One reason for this could be that they used fewer manipulations than the proficient explorers, but spent more time thinking about the problems (696 s). If we focused only on their test-level achievement, we would say they are medium problem-solvers; however, based on their behaviour at the end of the test, they became expert explorers.

Finally, we identified the thinking profile of the students whose exploration strategy was moderately successful on the easiest problems, but who did not know what to do on the more complex ones. They generally only spent a little time and applied a few manipulations to explore the problem environment, a tendency which could also have caused the very low performance (Mean = 18.9%, SD = 14.6).

These results also indicate and confirm the finding in the literature (e.g. Greiff et al. 2016) that investing time and effort in cognitive activities might be beneficial in problem-solving.

These three thinking profiles have been identified among primary students by [13] and [6], among others. They identified three more classes: non-performers, low performers and slow learners. In the present study, the sample was built from students entering the university,that is, the students were strongly selected for university entry. This may be why these groups dropped out of the analysis.

Based on the last research question, we separately tested the effect of the analysed process indicators on the latent profiles to check if any of the relations identified are invariant across the different profiles. All four process indicators proved to be non-invariant across the different profiles; that is, the number of manipulations made, time-on-task, type of thinking and type of strategy applied in the exploration phase influenced students' achievement on a different level depending on their thinking profile. Generally, all four indicators proved to be the most influential among the low achievers, but time-on-task and the change in number of clicks were not linearly dependent on the expertise level of the students. The strongest correlation was observed among the low performers and high performers, and a lower correlation was found among the rapid learners. The correlations between the manipulations made and time-on-task or number of clicks differed the most, indicating the very important role of time spent and number of trials executed in the exploration phase. This can be interpreted in two ways: the students in this group were undermotivated, almost clicking through the entire test without thinking, thus resulting in low achievement or – in correlation with this – they did not seek understanding. However, if they could have spent more time and effort on the exploration, they would have been more successful in both their exploration and problem-solving behaviour. That is, if we could have them spend more time with the problems and use more manipulations, they could have higher achievement.

9.10 Conclusions and Limitations

The present paper has explored the possibilities of using log data analysis in an educational context to make thinking and learning visible and to gain a better insight into students' thinking in a dynamic problem-solving environment. The study introduced how log data and learning analytics in technology-based assessment and training have entered the mainstream in the last ten years, thus opening new possibilities, such as providing the opportunity to raise research questions which could not have been answered with traditional assessment and analysis methods. Log file analysis can provide more elaborate feedback and even enables us to quantify qualitative differences. The second part of the paper offered an example and included a discussion of how log file data can contribute to a significantly better understanding of the phenomenon under examination and how students' qualitatively different thinking patterns can be classified and made visible.

A new labelling and scoring method was introduced for the log data collected during the exploration phase for dynamic problems developed with the MicroDyn approach. It was found that involving log data and using this labelling and scoring method in the analysis improve the reliability of the test.

It was also found that the use of a theoretically effective exploration strategy does not always result in high performance. It is assumed that students showing systematic strategies but failing to solve the problem may possess combinatorial skills but lack the necessary level of inductive reasoning. However, this calls for further analysis. It is more difficult to find an explanation for the phenomenon of students solving the problem without a right and effective strategy. Thus, solving the problem is not equivalent to using a proper strategy for identifying and collecting all the necessary information about the problem; a correct solution requires something more.

Using latent class analysis, three qualitatively different class profiles were identified at university level: (1) proficient strategy users who consistently employed optimal exploration strategies almost from the very first problem; (2) rapid learners who started out as low to intermediate performers in their exploration behaviour on the first, easiest problems, showed a rapid learning curve afterwards, began to use the partially isolated variation strategy increasingly and then the fully isolated variation strategy, and, by the end of the test, approached the high level of exploration behaviour of the proficient explorers; and (3) intermediate performers on the easiest problems, but low performers on the complex ones. No so-called intermediate strategy users were identified, i.e. students who used the partially isolated variation strategy almost exclusively on the test. The four process indicators tested proved to be non-invariant across the different latent profiles. That is, it is not sufficient to run analyses on the sample level; there are big differences between students within the different thinking profiles in the role of the number of manipulations made, time-on-task, type of thinking and type of strategy applied in actual problem-solving achievement.

One limitation of the study is the semantic embedding (i.e. fictional cover stories and variable labels) with the aim of minimizing the uncontrollable effects of prior knowledge. However, it is assumed that the semantic context will have an impact on exploration behaviour, that is, on how students manipulate and interact with the systems.

Acknowledgements This study has been implemented with the support provided from the National Research, Development and Innovation Fund of Hungary, financed under the OTKA K135727 funding scheme.

References

1. O.O. Adesope, A.G. Rud, Maximizing the affordances of contemporary technologies in education: Promises and possibilities, in *Contemporary technologies in education*, eds. by O.O. Adesope, A.G. Rud (Springer Nature, Cham, 2019), pp. 1–16
2. H. Aldowah, H. Al-Samarraie, W.M. Fauzy, Educational data mining and learning analytics for 21st century higher education: A review and synthesis. Telematics Inform. **37**, 13–49 (2019)
3. C. Angeli, N. Valanides, Using educational data mining methods to assess field-dependent and field-independent learners' complex problem solving. Education Tech. Research Dev. **61**(3), 521–548 (2013)
4. B. Csapó, G. Molnár, Online diagnostic assessment in support of personalized teaching and learning: The eDia system. Front. Psychol. **10**, 1522 (2019). https://doi.org/10.3389/fpsyg.2019.01522
5. X. Du, J. Yang, B.E. Shelton, J.L. Hung, M. Zhang, A systematic meta-review and analysis of learning analytics research. Behav Inform Technol 1–14 (2019)
6. S. Greiff, G. Molnár, R. Martin, J. Zimmermann, B. Csapó, Students' exploration strategies in computer-simulated complex problem environments: A latent class approach. Comput. Educ. **126**(11), 248–263 (2018)
7. S. Greiff, S. Wüstenberg, F. Avvisati, Computer-generated log-file analyses as a window into students' minds? A showcase study based on the PISA 2012. Comput. Educ. **91**, 92–105 (2015)
8. J. Hattie, *Visible Learning: A Synthesis of over 800 Meta-Analyses Relating to Achievement* (Routledge, London, 2009)
9. L. Johnson, S. Adams Becker, M. Cummins, V. Estrada, A. Freeman, C. Hall, *NMC Horizon Report: 2016 Higher Education Edition*. (The New Media Consortium, Austin, Texas, 2016). Accessed from: http://cdn.nmc.org/media/2016-nmc-horizon-report-he-EN.pdf
10. Journal of Learning Analytics. Solar Society for Learning Analytics Research. https://epress.lib.uts.edu.au/journals/index.php/JLA/about/editorialPolicies#focusAndScope
11. C. Lotz, R. Scherer, S. Greiff, J.R. Sparfeldt, Intelligence in action–Effective strategic behaviors while solving complex problems. Intelligence **64**, 98–112 (2017)
12. T. Martin, B. Sherin, Learning analytics and computational techniques for detecting and evaluating patterns in learning: An introduction to the special issue. Journal of the Learning Sciences **22**(4), 511–520 (2013)
13. G. Molnár, B. Csapó, Exploration and learning strategies in an interactive problem-solving environment at the beginning of higher education studies, in *Knowledge Management in the 21st Century: Resilience, Creativity and Co-creation. Proceedings IFKAD2017*, eds. by J.C. Spender, T. Gavrilova, G. Schiuma (St Petersburg University, St. Petersburg, 2017), pp. 283–292
14. G. Molnár, B. Csapó, The efficacy and development of students' problem-solving strategies during compulsory schooling: Logfile analyses. Front. Psychol. **9**, 302 (2018)
15. G. Molnár, B. Csapó, Making the psychological dimension of learning visible: Using technology-based assessment to monitor students' cognitive development. Front. Psychol. **10**, 1368 (2019a)
16. G. Molnár, B. Csapó, How to make learning visible through technology: The eDia-Online Diagnostic Assessment System, in *CSEDU 2019. Proceedings of the 11th International Conference on Computer Supported Education*, vol. 2, eds. by H. Lane, S. Zvacek, J. Uhomoibhi (Scitepress, Heraklion, Greece, 2019b), pp. 122–131
17. G. Molnár, S. Greiff, S. Wüstenberg, A. Fischer, Empirical study of computer-based assessment of domain-general dynamic problem solving skills, in *The nature of problem solving: Using research to inspire 21st century learning*. ed. by B. Csapó, J. Funke (OECD, Paris, 2017), pp. 123–143
18. N. Pachler, C. Daly, Y. Mor, H. Mellar, Formative e-assessment: Practitioner cases. Comput. Educ. **54**(3), 715–721 (2010)
19. K. Tóth, H. Rölke, F. Goldhammer, I. Barkow, Educational process mining: New possibilities for understanding students' problem-solving skills, in *The Nature of Problem Solving: Using*

Research to Inspire 21st Century Learning. ed. by B. Csapó, J. Funke (OECD, Paris, 2017), pp. 193–209
20. O. Viberg, M. Hatakka, O. Bälter, A. Mavroudi, The current landscape of learning analytics in higher education. Comput. Hum. Behav. **89**, 98–110 (2018)
21. R. Vollmeyer, B.D. Burns, K.J. Holyoak, The impact of goal specificity on strategy use and the acquisition of problem structure. Cogn. Sci. **20**, 75–100 (1996)
22. R. Vuorikari, JC Muñoz, Research evidence on the use of learning analytics: Implications for education policy. *JRC Science for Policy Report*. (European Commission, 2016). Accessed from https://publications.jrc.ec.europa.eu/repository/bitstream/JRC104031/lfna28294enn.pdf
23. A.F. Wise, Learning analytics: Using data-informed decision-making to improve teaching and learning, in O.O. Adesope, A.G. Rud (eds.), *Contemporary Technologies in Education* (Springer Nature, Cham, 2019), pp. 119–144
24. N. Zoanetti, P. Griffin, Log-file data as indicators for problem-solving processes, in B. Csapó, J. Funke (eds.), *The Nature of Problem Solving: Using Research to Inspire 21st Century Learning* (OECD, Paris, 2017), pp. 177–192. Accessed from https://doi.org/10.1787/9789264273955-13-en

Chapter 10
AI in Consumer Behavior

Dimitris C. Gkikas and Prokopis K. Theodoridis

Abstract E-commerce is one of the fastest changing industries and consumers try to purchase most of the goods online. Following the internet revolution, retail and promotion, daily data generation led to a point where marketers could not further process, using the traditional statistical ways, the new demands of vast data volume. The use of big data analytics emerged, and the rise of AI came along with predictive analytics including machine learning and data mining solutions. Artificial Intelligence has both evolved and affected the way marketing is implemented. AI is proven to be a sophisticated way of analyzing and processing data as well as decision making. The privilege of its use is considered an asset to the business executives who understand its potential. Considering AI as a tool, it may provide massive data processing and prediction accuracy. Referring to marketing science, marketers have witnessed the benefits of AI by predicting consumers behavior. Since consumers behavior varies, brands struggle for customer satisfaction, they invest time and money to highlight their products or services potentials, better define their market share, and classify customers' needs. Marketers have been in pursuit of customer satisfaction using AI tools to read web metrics and optimize reach and conversions strategies. Machine learning, natural language processing, expert systems, voice, vision, planning, and robotics are the main AI branches that companies use to stay ahead of the competition. Business objectives aim to attract new customers, predict consumer behavior, along with the capability to personalize and predict demand using "smart" systems which allow them to increase sales, mitigate the decision-making risk and increase customer satisfaction, customer loyalty, and sales predictions. This chapter tries to map the stages of AI contribution to consumer behavior describing the essential marketing milestones.

Keywords Artificial intelligence · Consumer behavior · Digital marketing · Machine learning · Decision making

D. C. Gkikas (✉) · P. K. Theodoridis
University of Patras, 2 Seferi G. Str., 30100 Agrinio, Greece
e-mail: dgkikas@upatras.gr

M. Virvou et al. (eds.), *Advances in Artificial Intelligence-based Technologies*,
Learning and Analytics in Intelligent Systems 22,
https://doi.org/10.1007/978-3-030-80571-5_10

10.1 Introduction

The fast growth of information technology has generated new potential in marketing science predictive significance, effectiveness, and performance. In order companies to gain such business competitive advantage and conduct effective marketing strategies, are essential to gather and store consumer information that will help analyze, model, personalize, predict, and influence their potential customers behavior and intention to purchase products or services. Due to technological progress, people's lives have changed along with their purchasing behavior. Societies are constantly changing and evolving, they are improving their structures, communications, and working environment in daily basis. Consumers' habits change and differ accordingly, based on a set of factors that seem to affect consumer choices and reactions. Such factors are considered to be cultural, social, financial, geographical, political and tend to affect customers behavior when coming to decide to purchase a service or a product [1].

Electronic commerce has given companies the opportunity to sell their products or services through their websites or e-shops to their audience. The number of transactions and behavior data generated from those purchases give these companies the advantage to understand the way that consumers think and act during their consuming journey. There is no comparison to the recent past when people used to visit a website and perform a purchase without the intervention of other intermediaries such as recommendations, and bestsellers. However, from the retailers' perspective, online purchasing has entirely changed. Similarly, the consumers' behavior and decision making have also altered. Reviews, comments, prices, site-speed, button size, colors, payment methods product specifications, shipping time and product comparison among websites are some of the factors that seem to affect the majority of consumers. Perhaps having less choices and the least amount of information could make easier for marketers, analysts, or retailers to predict what the customers would like to buy, but with the current volume of daily generated data and the amount of information provided through the marketing channels any predictions or even suggestions seem impossible. Considering that amount of data, it could be impossible, even for the entire company's personnel, to be able to manipulate and handle customers data and provide competitive suggestion using obsolete statistical ways [2]. Consumer satisfaction through consumer behavior and user experience define the level of success. Sophisticated tools powered by artificial intelligence are required to reveal customer purchasing patterns and turn consumer decision making and predicting into simple decision making for marketers, retailers, and companies [3].

Before artificial intelligence emerged, digital marketing took place and still affects consumers' journey. Digital marketing or online marketing defined as the marketing of products or services using information technology through digital channels such as search engines, social media, email, and other websites to connect with current and prospective customers. Digital marketing has changed the way brands and businesses use technology for marketing using techniques such as search engine optimization, search engine marketing, influence marketing, content marketing, chatbots etc. [4].

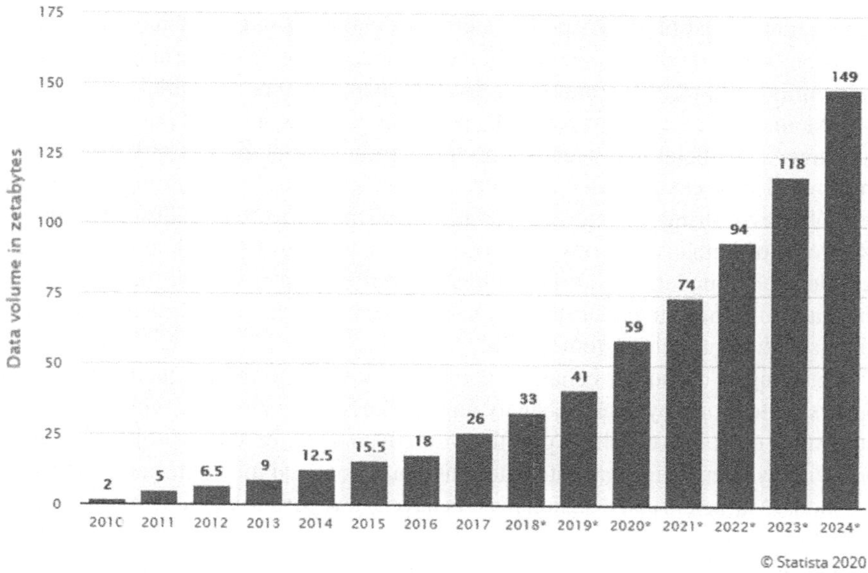

Fig. 10.1 Worldwide Data Created from 2010 to 2024. *Source* https://www.statista.com/statistics/871513/worldwide-data-created

As digital platforms become "smarter" and more widely used by retailers and as consumers use digital devices instead of visiting physical shops, digital marketing applications have become important and efficient [5]. Worldwide users are constantly connected to digital platforms daily generating enormous volume of data called big data. It is estimated that by 2024 the total amount of data created, captured, copied, and consumed in the world is forecast to increase rapidly, reaching 149 zettabytes (Fig. 10.1).

Big data analytics refer to gathering, analyzing, and providing potential correlations or causations among certain variables or attributes in order to reveal patterns and hidden information that will help decision makers make more precise, affordable, promising and liable decisions. Nevertheless, big data analytics are based on predictive analytics using machine learning and data mining known as sophisticated artificial intelligent algorithms [6].

Artificial intelligence (AI) provides marketing science the ability to efficiently handle big volumes of daily generated data and overcome human's inability to process these data using obsolete ways providing highly accurate results. Brands need to record, and research consumers' journey in order to predict consumers' behavior. Knowing consumers next moves, marketers are able to make more efficient decisions through automated "smart" procedures and platforms. The true potential of artificial intelligence is showed through the results of big data, machine learning and data mining applications in the companies. As previously mentioned, as marketers managed to model, find patterns and manipulate unprecedented scale data it gave them the advantage to eliminate mistakes, narrow the targeted groups of consumers, and target the proper audiences using the most suitable medium per occasion. When it comes to prediction, machine learning, deep learning, along with data mining and results extraction it gives decision makers a significant advantage to predefine and accurately predict the potential needs, trends, and purchases by the consumers. Artificial intelligence algorithms train their models in order to be self-improved by processing huge amounts of data. Since machine learning can handle big volumes of strings of data, they certainly have the ability to partially "understand" the world of humans in terms of choices and needs. However, while many brands rebuild their foundations on innovation and artificial intelligence, there are many that have to manage to handle the increasing volume of data which can improve their understanding of their customers [7].

Referring to consumer behavior, computers are able to predict the behavior and buyers' intentions by storing, analyzing, processing, patterning, identifying and generating data. These abilities give them the advantage to be able to project consumer behavior by interpreting data into emotions and actions and substitute human communications. Marketers and decision makers have a new power in their hands with huge potentials which enables them to perform customer segmentation and apply the insights to develop customer engagement creating a new customer value [6].

Despite the advantages of this technology and its potentials there are some important ethical matters emerging which society must face when it comes to personal data and people privacy. Building the right governance structures and making sure ethical principles are intact is essential. What the barriers of artificial intelligence are, who accesses this data, what sort of personalization artificial intelligence can get are some of the questions that must be thoroughly investigated. This chapter tries to provide a logical cohesion among the aspects of consumer behavior, marketing, e-commerce, and artificial intelligence and how they utterly end up combined aiming to provide new knowledge.

10.2 Literature Review

In this work, the authors identify critical information about the potential of AI application through the path of several marketing aspects towards explaining its contribution in consumer behavior. Hence, to the best of their knowledge the authors have mapped the most representative literature works referring to these fields of study which reveal the significance of using technology in consumer behavior analysis. Nevertheless, referring to consumer behavior analysis the area of research interest is vast, and the need of future scientific works is essential.

Referring to decision journey stages, the authors of this work state the most significant research papers of how business indicators such as consumer decision journey and marketing using rule-based and machine learning classifiers can be automatically classified [8, 9]. In terms of generating a consumers' behavioral model from observed consumers behaviors and attitude in a given competitive market frame, there have been some serious scientific attempts integrating concepts modeled by various fields of study including behavioral attitude, decision making and data mining techniques [10–12, 80].

Nowadays, customer behavior model's generation are usually based on customer data analysis using data mining algorithms. However, this is a difficult task for researchers because the risk of misinterpreted or biased data and results may mislead the outcome of the research. Thus, applying data mining techniques on customer behavior models requires the right approach. Nevertheless, several studies reveal how consumer behavior prediction using data mining techniques was successful in developing association rule mining models to predict customer behavior managing to export new sets of rules [6, 13, 14]. Compared to similar consumer behavior prediction attempts there is one which demonstrates an alternative use tree-based feature transformation and machine learning algorithms, instead of using ensemble algorithms, a simple algorithm is used to predict purchase behavior based on transformed features [15]. Another approach in predicting consumers' behavior this time using random forest data mining technique proved to be effective since it has managed to succeed high accurate results [16].

In the same spectrum of consumer behavior prediction there is a substantial number of studies which refer to in-store behavior monitoring using eye tracking and embedded sensors technology. Different processes like scene perception, reading, visual attention and decision making can perform high accuracy prediction. Factor such as focus time, positioning, shopper motion, total fixation duration, fixation count affect consumer's final decision [17–21].

Mobiles have a significant impact on redefining the consumer behavior models by altering the factors such as access location, browser type, time, mobile operating system that affect behavior [22–24]. An undeniable fact is that, like mobiles, social networks are powerful tools in the hands of marketers that can predict or affect purchases and consumer behavior. Online businesses are able to communicate with their customers at any time or place around the world. Companies realized the potential of using social media platforms to influence their customers, embedding social media marketing approaches in their overall marketing strategies to benefit their businesses [25]. This is a new reality that emerges, and marketers must adapt to these changes. The amount of daily generated raw data is an opportunity for companies to apply data mining techniques and extract new knowledge about their customer behavior and intentions [26].

10.3 Artificial Intelligence (AI) in Consumer Behavior

In a world where billions of transactions take place electronically, consumer behavior passes through artificial intelligence. In order for someone to conceive the role of artificial intelligence in consumers' journey, including buyers' decision making and consumers behavior, there is a logical path which must follow. It is essential to have an understanding of the basic principles of AI and how it is affecting certain customers' journey phases. What AI is, how it affects marketing science and e-commerce, what

AI marketing is, and finally how AI entirely affects and get affected by consumers' behavior and technology are thoroughly explained.

10.3.1 Artificial Intelligence

10.3.1.1 AI Theory

MIT Technology Review reports that, artificial intelligence simply refers to "machines being able to learn, reason, and act for themselves." [27].

Artificial intelligence (AI) refers to the computer's ability to perform calculations in a similar way with living beings which has a certain level of intelligence. It is a combination of many different technologies aligned together to enable computers to sense, understand, act, react, and learn like humans including activities like reasoning, meaning, generalizing and recursive learning. AI consists of techniques like machine learning, deep learning, data mining and natural language processing. Thanks to these techniques AI applications can train their models, analyze new information and provide highly accurate results [28].

However, even though it has been proven that computers can be programmed to undertake complex tasks with great success despite the constant progress in computer processing power, no programs can be yet found to compete with human way of thinking [29].

Nevertheless, there are some software tools which have comprised the requirements to match human experts' levels in conducting certain calculations and provide high accurate results. Companies realized that they can be leveraged by possessing huge volumes of unstructured data which can set the foundation for future power and dominance [30].

In order to understand the significance and the scale of this data and how it can be patterned the use of five Vs of data theory which breaks data in five dimensions is essential: Volume, Velocity, Variety, Veracity and Value. From traffic patterns and music downloads to browsing history and medical records, data is recorder, stored and analyzed to enable the technology and services that the world relies on every moment.

The volume refers to the amount of big data instantly generated, such as keyword search queries on search engines, website clicks or browsing [31].

The velocity refers to the time or speed in which big data can be processed, such as real-time demands on the web, personalized advertisements, email automation, call data processing, product recommendations, or weather data.

The variety refers to the sources' types from where the structured or unstructured big data came from, such us social media data, text, sensor data, audio, video, research data or website data. Veracity deals with the validity and the trust towards the big data, such as self-reported data versus data gathered automatically.

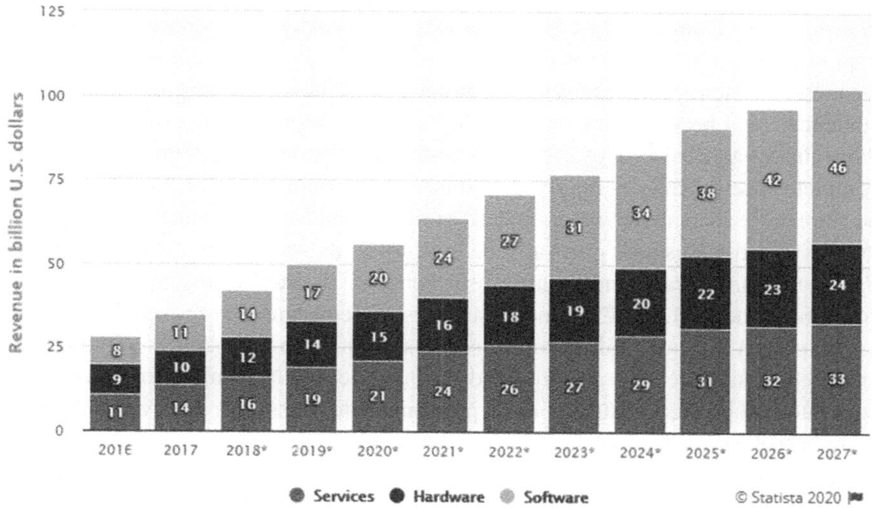

Fig. 10.2 Big Data Revenues from 2016 to 2024. *Source* https://www.statista.com/statistics/301566/big-data-factory-revenue-by-type

The value refers to what is the added value of the data gathered that can contribute to the process [31]. The revenue from big data is estimated to be 103 Billion USD in 2027 (Fig. 10.2).

Such solutions have already been applied to businesses helping them lead innovation and have the competitive advantage by using raw data.

After data is gathered, analysed and processed it produce suggestions that help marketers to make prosperous decisions, segment their audiences, predict consumer behavior, and identify market trends [29].

Companies which have acquired and embraced AI have tripled the return on their AI investments by embracing the use of systems that are able to self-train, learn and evolve [30].

10.3.1.2 AI Advantages

Artificial intelligence can be used in several aspects of peoples' daily lives. However, the best way to examine the use of AI is to define what are its capabilities and how they can leverage businesses. Applying AI to companies it helps them increase their value, stay ahead of the competition, and secure their future. AI contributes to cost reductions through efficient resource management and effective data handling. Due to the self-learning model it prevents system failures and disruptions by predicting best possible maintenance times. High accurate results produced by a series of machine learning techniques help decision makers make effective and qualitative decisions.

Humans do not have the mental ability to process big volumes of data simultaneously. This lack or processing capabilities generates gaps of hidden information and missing value. Computers are able to process and provide data reports which reveal opportunities, trends, risks or potential threats. This insightful capability leverage businesses which want to present new products, use new mediums, target narrowed audiences, and perform new business strategies in the market more rapidly. Under these circumstances' employees' role is empowered by dedicating time and efforts to complete specific tasks of highly importance in a certain period of time. Employees become more specialized, more objective oriented, and more productive minimizing the risk of error. Business goals also include customer satisfaction. Thus, it is essential to constantly track and monitor user behavior in order to instantly respond to consumers' needs [29]. Customers' services are constantly improved through the users' personalization and customer segmentation from data that is gathered and analyzed. Insight reports are provided by the machine learning algorithms which continuously process customer data, purchase data, consumer behavior data on a daily basis [30].

Apparently, AI can be used in many ways to help businesses grow. Hence, it is essential business owners to determine what their business needs and objectives are and what would be the appropriate AI tools to serve these needs. Business so far are skeptical on how they could apply AI in their current structures. Hence, it is essential to define their needs and values in order to examine the potential use of "smart" systems [29]. Business value, services, customers' satisfaction, future investments, future demand are some of the factors that will affect the AI scaling.

Society should examine without suspicion the use of "smart" systems and avoid exaggerations. Waiting for society to technologically mature and the business world to understand the true effect of AI in industry, the doubts and the assertion about AI concerning a potential manpower reduction in industries has started fading.

In this AI journey of acceptance, international legislation, government rules and ethical frameworks should be applied into the development of algorithms and software in order to protect people's privacy and businesses strategical assets [30].

10.3.1.3 AI Marketing

Following the application of artificial intelligence in marketing science, a new term of marketing emerged called AI Marketing. Forbes magazine takes a high-level view of AI marketing, describing it as "a method of leveraging technology to improve the customers' journey." AI could be simply explained as a system that conducts complex calculations using the way that humans use to think but at the same time encompasses systems which can analyze massive amounts of data and find compelling patterns which humans find difficult to process. These tasks include problem-solving, patterning and recognizing emotions, predicting, imitating, and even diagnosing [32].

AI marketing could be defined as the process of using artificial intelligence as a tool to automate procedures like data collection, data analysis, data process, giving decision makers the ability to empower decision making processes and transform them to enhanced consumer-centric decision-making mechanisms.

AI optimizes the data analysis process delivering on-time and on-demand insights improving the user satisfaction and conversion rate without any marketers' assistance. AI marketing consists of three main pillars: Big data, machine learning and personalized marketing services [33].

Nowadays, most companies make use of AI techniques like advanced analytics and machine learning trying to create patterns that will leverage their marketing strategy [34].

AI presence has generated new opportunities for businesses, especially in the digital marketing and e-commerce sector. There is a great potential of AI use in marketing science and consequently in marketing business. AI will transform the marketing scenery by improving most of the marketing sections. Advanced big data analytical tools, sophisticated machine learning algorithms, big records of stored data, and intelligent agents lead to an unprecedented technological progress along with user's satisfaction and multilevel implications across the marketing industry.

The customer's journey will be benefited by such technological progress and consumers will be leveraged using it. Some of the insights that will create efficient consumer touchpoints are content curation, content marketing, data analytics, machine learning, audience targeting, personalized, customer searches and recommendations, enhanced digital advertising, chatbots, behavior analysis, predictive analytics, email automation, social media marketing, website testing, ai-driven sales prediction, user experience (UX/UI), fake reviews filtering etc. [32].

AI is able to optimize marketing campaigns, moderate the risk of human error, decrease much of guessing in customer interactions. However, in order for businesses to connect with their customers the human presence is essential.

Empathy, compassion, human touch, customer support and storytelling cannot yet be performed by computers and probably never will. At the end of the day, AI is not bound by human limitations [32].

10.3.1.4 AI in Content Marketing

Referring to content marketing, AI has boosted content creation and curation. Bots scan the internet to find relevant content that can connect with visitors on websites. AI has powered content marketing to a certain point where marketers are able to map out an end-to-end content marketing strategy to use it for consumer reach and engagement [35]. AI has proven to be capable in providing content generation and curation to targeted audiences that will allow increased engagement with customers. Therefore, AI technology has already automated content creation been able to provide articles, topics, reports, translations, etc. [32].

Content generation and curation has reached to a point where it is difficult for someone to tell the difference whether that content was written by a human or generated by a machine. Undoubtedly, computers have reached to certain level of trustworthiness in terms of content generation. Computers, thanks to natural language processing (NLP) are able to read, interpret, pattern and understand text and write reports and news and content based on data [36]. Content marketing is constantly evolving, providing customer-oriented articles, human engaging web pages, and personalized communication. Companies take advantages of this technological "blooming" and use it to increase customer targeting, satisfaction, and conversion. Living in times of a multilevel interdisciplinary content storming the need of authenticated, targeted, and engaging content has emerged. Content marketing, apart from content creation, deals also with the direct customer experience of satisfaction and help along customers' journey. Satisfied customers will lead to increased customer engagement, high conversion, and re-engagement rates [33]. Marketers' problems lie in how they will establish a win–win relationship with the consumers. Through companies' communication to the right audiences, they will make consumers more receptive to their services and products. Listening to what customers have to say and paying attention on negative reviews, feedback and provide solutions to their problems, in times when information has overwhelmed the internet, will provide customers the feeling of care and safety. With this information, decision makers will be able to perform a marketing strategy which will target relevant audience in order to influence potential purchasing decisions. Being able to target the right audiences, it increases the conversion rate and revenues and builds relationships with the current or potential customers by creating and personalized quality content [35].

10.3.1.5 AI in Content Personalization

In a world of excessive internet usage with countless options, products, companies and opinions, decision makers aim to create personalized content for online customers. Matching content to user queries or interests is essential when potential customers seek particular information, product or service in a website, and should this take a length of time, they are likely to leave the webpage. Nowadays, AI has offered a way in solving these situations by helping customers to find what they are seeking for in a much more natural way. Content personalization using AI has become one of the most efficient ways to curate stored or real time web content or provide product recommendations to web visitors [37, 82].

The rise of the smart devices and messaging applications, combined with the rise in AI, provides the opportunity for companies to restore communication with their customers and be more proactive. Customers need to feel unique and it is important for companies to invest time and money to build strong relationships with the customers by providing them with a top-quality customer experience [83]. These conversations shape the future relationships between companies and customers and the extracted data reveal consumer behavior towards of better understanding and proactive caring [33].

The content marketing campaigns can be optimized and reduced to fit to specific audience avoiding useless information effusion increasing the return of investment (ROI) at the same time. The key towards success is the consumers' analysis, segmentation and targeting in order for answering their needs. Only then can decision makers assume that content marketing strategy performs well. When companies fail to engage customers through content marketing is because they do not focus on the consumers' needs. Thus, they do not manage to increase the level of ROI.

There are certain requirements for companies which need to generate website traffic and user engagement. These requirements include goals setting, focus on the right people, on the right timing using optimized engaging content. To begin with, goals and objectives identification is essential for the entire strategy to be completed successfully. Goals may include potential new customer churns or website organic or paid traffic, visiting time, clicks, revisits, conversions etc. Following, the use of certain analytics is imperative to segment the audience on time. Hence, the use of geographical, demographical, user behavioral, real time data etc. can help distinguish groups of customers according to their needs, place, time, and behavior [38].

Finally, articles, topics and information that answer consumers queries is essential to be generated. The companies must be able to answer any information their clients seek. Customer engagement must be constantly recorded and analyzed to improve content performance [33].

AI gives the ability to monitor performance insights that helps to determine the overall company marketing strategy and being more energetic when it comes to content scaling [35]. Scaling up targeted content distribution can be especially challenging. Instead of manually writing algorithms or simply guessing what customers would like to see, machine learning and intelligent analytics can be used to provide up-to-date and personalized content, and to scale this to the vast amount of touch points, channels, brands and countries. Brands can more effectively and precisely predict customer behavior and respectively offer recommendations [39, 82].

10.3.1.6 AI in Digital Advertising

Companies today have started to commercially use artificial intelligence tools to perform customer analysis. They identify and segment audiences, build advertisements campaigns, test alternatives, improve efficiency, and optimize advertisements spend in real-time, and at scale. Artificial Intelligence affects search engine marketing and the way that companies use online advertisements. Nowadays, digital advertisements are data-driven using machine learning algorithms to read, analyze, track, target, and optimize consumers and results. AI-based delivery systems using machine learning, data mining and big data, segment and personalize advertisements accordingly showing the proper advertisements in front of the right audiences [40]. The scaling and the audience analysis follow the same logic as content marketing use. Advertisements' development is not just a matter of creativity but also a matter of strong data collection, analysis, and mining that companies must be aware of to increase customer satisfaction and ROI. AI use on advertisements refers to the advertisement platforms, performance and spend optimization and advertisement creation [76]. "Smart" advertisements are about targeting and demonstrating the right ad to the right audience. Unfortunately, consumers are daily bombarded with irrelevant ads. Most of them become annoying but for the companies it has several significant drawbacks, it creates bad reputation, bad brand awareness and increases the

CPM advertising costs for not any considerable results. When this situation occurs, marketers, advertisers and brands lose money and brand loyalty. However, brands can easily increase the level of ROI by using AI tools for advertisements and automatically place their advertisements to relevant audiences. AI technology can assist brands and decision makers to strategically identify trends, predict opportunities or threats keep at the same time a low ratio between money and value [32].

10.3.1.7 AI in Chatbots

New-age marketing requires a new-age communication medium that should quickly, systematically, and accurately embody real time correspondence with the users. Hence, chatbots are the type of AI-enhanced systems that can greatly impact customer experience. Chatbots are already being used in most websites around the world answering users frequently asked questions. A chatbot mainly consists of an algorithm that stores and answers certain questions along with their answers and communicates with the user through a simple messaging interface. After social media platforms, chatbots are considered to be the future in communication between consumers and marketers [36, 41]. Chatbots allow businesses to handle requests in high peak times during the day as they operate 24/7. They can help businesses by decrease response time, freeing up employees from answering routine questions and focus on their customers. Nonetheless, chatbot mainly serve this purpose and technology works as a substitute of human presence where primarily customer service personnel should handle customer issues in person rather than machines. However, due to its sophisticated software AI-enhanced chatbots make a great difference to customer experience. "Smart" chatbots have the ability and the "intelligence" to real-time create content and communicate with users using, genuine curated responses [32, 42]. As the use of AI technology increases and people become more familiar with the new technology the use of chatbots tends to become irreplaceable due to fact that are called to respond to both simple and complicated questions. Chatbot give marketers the opportunity to serve their clientele offering them a pleasant, practical, and valuable experience [37].

Companies are estimated to be able to save around 8 billion USD annually by 2022 (Fig. 10.3) from an evolving technology which effectively supports and enhances human–machine communication providing answers, questions, and recommendations [44].

10.3.1.8 AI in Voice Search

Voice and visual search consist of what is called "sensory search", including text search by users. Voice search refers to natural language processing technology (NLP) which incorporates artificial intelligent algorithms to identify, record, transform it to text, text mining, analyze, pattern, and interpret human voice in order to deliver results when voice commands are given. Even search engine optimization for websites has

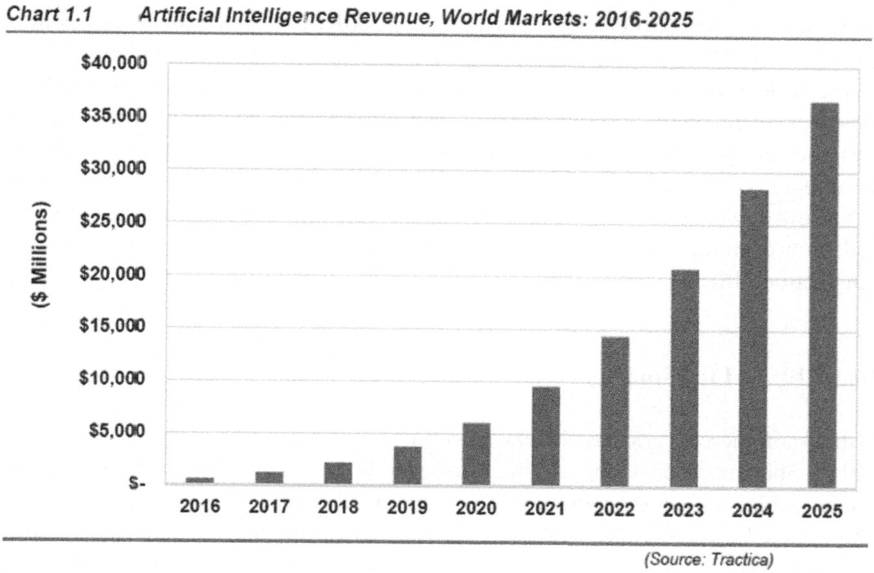

Fig. 10.3 AI Revenues in USD in World Markets from 2016 to 2025. *Source* https://tractica.omdia.com/newsroom/press-releases/artificial-intelligence-software-revenue-to-reach-59-8-billion-worldwide-by-2025

altered and search engines require change of plans concerning the short or long-tail keywords in order to match with the future voice searches [45]. There are already numerous voice search applications therefore, the marketers acknowledge voice search significance and that will be widely used for searching. From a marketing perspective voice search will hugely impact the marketing sector. Millions of voice searches are performed every day hence, machine learning algorithms have the data required to evolve reaching to a point where understand meanings. The searches will be transformed from simple or long-tail keywords commands to conversations between human and computer. Instead of writing people might prefer to use voice commands or voice searches which will save them time and effort [36].

10.3.1.9 AI in Visual Search

Visual and voice search are strongly related with artificial intelligence. Visual search help people search for information which is visualized. Using AI-powered image detection software, users are in position to identify a real object or an object in a picture and provide information [46].

Visual search platforms vary. There are platforms which give consumers the opportunity to take pictures using their mobile phones and in return provide them with information or similar photos. There are also platforms which give customers options to select similar photos to those they search for in order to provide them suggestions. Finally, users can directly conduct image searches instead of text searches [47].

Home decoration and fashion industries are currently benefited by this technology. However, there seems to be a problem. Not all searched end up successfully and customer are less than satisfied which increase customer dissatisfaction. Thus, to avoid poor user experience visual search also allows users to perform voice search to narrow down, specify and improve search results, to find and compare products and allow customers to conduct comparisons among similar product to increase user satisfaction [48].

10.3.1.10 AI in E-mail

In the e-commerce world, customers are getting bombarded with emails, most of them end up unopened consisting a sort of noise to the recipient. Due to AI technology it is possible, instead of spamming, to automatically generate scalable individually personalized curated content marketing emails where the user receives information according to his interests. In comparison to mass emailing, AI has performed better by providing the ability to enable decision makers fairly treat and reach their audiences by identify and analyze new segments, optimize messages using conducting keyword research and dynamically perform marketing strategy [81]. Through specific email marketing campaigns with personalized content the targeted audiences would be interested to receive similar information based on past subscriptions, geographical, demographical, behavioral, or real-time data [37].

Artificial intelligence impacts email marketing in several ways but mostly it helps increase revenues using AI-tools for email marketing and decreases costs by generating easier to read, qualitative and more engaging content campaigns [49]. AI contributes to email content performance optimization using two ways. To begin with, due to previews email data provided to train the natural language processing (NLP) model the AI powered tools generate email subject lines better in quality than the actual users. The generated emails' content performs better in opening rates and revenues. Finally, NLP models manage to personalize the content down to the email recipient. This level of personalization resonates with personalized user products or services recommendations and targets individuals aiming to increase user engagement and conversion rates [49].

10.3.1.11 AI in Social Media

Companies using artificial intelligence application on social media platforms, collect substantial amounts of data from their customers with regard to their preferences, what content they share, what time they share it and some demographical and geographical insights. Whether companies use social media analytics to segment their audiences and measure the impact of their products or services AI takes it a step further by using AI algorithms. Social listening, sentiment analysis, text mining and a behavioral data are stored and processed to provide new set of customer behavior patterns [50].

Using AI can help marketers generate authentic content for social media sharing and distribution. Copywriters and social media analysts are able to faster create targeted social media content. Social media tools using AI can generate patterns of how to create and curate content from past posts' data and determine which colors, keywords, and hashtags they will use in order to increase user engagement. In addition, AI social media tools may provide predictions about what posts will increase reach, clicks, shares, conversions, or brand awareness in the future. Marketers can monitor, analyze audiences, generate content, and address their clients across many social media channels using powerful platforms for handling multiple advertisements from many social media channels accounts from a single location [51].

Referring to predictions and insights "smart" social media tools outperform the basic social media business services by delivering a series of advanced services. AI can easily identify which channels will have the most significant reach and which audiences will be most receptive. These services include content curation, user profiles segmentation, audience targeting, market share increase, consumer trends detections, optimal social media time scheduling [77].

Apart from social media organic posts there are also the paid posts which are powered up since AI manage to generate advertisements automatically aiming for impression increase, clicks performance, conversion optimization and user engagement. The characteristics mentioned including targeting, advertising effectiveness, and creativity can help to achieve paid results beyond expectations [52]. Referring to advertisements, AI can analyze PPC advertising across platforms and offer recommendations for maximum performance [53]. These new features can assist marketing experts scale the customer churn by monitoring behaviors, observing consumers discussions, and what they search for [40].

10.3.1.12 AI in Website Testing

Application of AI in marketing can also be found in website quality testing tools. Business reputation on the web relies on the quality of its web entities such as websites and social media business pages. Optimization can include development, design, interaction, speed, and content. An optimized website will convince customers to engage [37].

In order for a company to define the optimal website in terms or conversion rate potentials several websites are tested. The testing methodology, known as A/B Testing, deals with structural performance instead of customers using experience. Hence, the need of AI-driven tools emerged, and marketers are able to put thousands of websites designs under a multifactor examination to find that will increase conversion rates. These tools generate patterns that scale the level of conversion success and use them to predict better user experience, higher conversions rates and revenue [37].

10.3.1.13 AI in Reviews

Nowadays, fake reviews consider to be one of the major e-commerce drawbacks for shoppers and shops. Due to the fact that positive or negative reviews affect buyers' decision and usually fake reviews originate from competitors or bots, businesses managed to use AI tools to distinguish true from fake reviews [54].

Due to previews data reviews AI-powered tools generate patterns that are able to verify whether a review is verified and helpful and weight these answers accordingly as the most assistive and customer centric. Original reviews do not help customer to find a product but also protect brands reputation. Finally, other customers are able to mark as helpful some given reviews for their authenticity [54].

10.3.1.14 AI in Robotics

Robots consist a major factor that will affect customer behavior and decision in the near future. Brands have the chance to supply their customers by using "sky-rocket" technology like drones or autonomous machines. Automation seems to be "one-way street" to success and companies which will understand its value they will be propelled forward commercially from the competition. A growing number of technology-familiar customers demand speed and accuracy when they search or purchase a product or service. Product publishing, post scheduling, product delivering, sales scheduling, product recommendations, customer caring, visual assistance, or even physical nature services like coffee making are considered to be some of the robotic new-aged tasks towards a better consumer experience [54].

10.3.1.15 AI in Analytics

AI enhanced analytics or predictive analytics refer to data mining, statistics, and data modelling to make predictions. Since the daily generated data incrementally increase, predictive analytics will shape a new digital marketing landscape [36]. Professions like data analysts, web analysts, digital analysts, and developers are increasingly demanded by the companies. Since big data cannot be handled by humans alone the urge of AI usage emerges. Based on previous consumer behavior machine learning, deep learning, data mining can assist marketers to provide accurate customers predictions [32].

Artificial intelligence is integrated into every stage of data processing. With personalization, speed, and accuracy that AI offers it brings great value to the table of companies marketing plans [55].

Predictive analytics include project definition, data collection, data analysis, statistics, modeling, deployment, and model modeling. Nowadays, Predictive analytics apply customer relationship management, cross selling, data collection, heath care, risk management, fraud detection and direct marketing [55].

Predictive analytics help marketers acquire knowledge and potential to be ahead of the rest competition. In terms of user satisfaction and revenues user modeling and audience segmentation, lead scoring, pricing solutions, effective campaigns, and sophisticated recommendations systems can be game changers. There are certain types of platforms using predictive analytics such as social media analytics, automated decision systems, marketing analytics, and cross-departmental analytics [36, 56].

Social media analytics refer to the process of collecting and analyzing user behavioral data from social media platforms using some the audience analytics, performance analytics, competitive analytics, paid social media analytics, influencer analytics, sentiment analysis and customer service analytics which help marketers track uses' behavior, business pages, content performance and products performance [56]. Business marketing analytics refer to in scale audience segmentation, user behavior modeling, feedback analysis, needs customization and insights reposting, analysis, and suggestions [56].

Automated decision systems utilize data collection and data modeling to improve marketing operations performance by conducting user personalization and customer-oriented marketing efforts [33]. Cross-departmental analytics deal with independent, loose-coupled, and multilevel access to business database analytics from different departments in order to conduct a certain data research [33].

10.3.1.16 AI in User Experience (UX)

According to the authors of this work, probably the most significant part of the consumers journey was deliberately left to be explained at this point of the chapter where user experience (UX) meets consumers behavior. Due to the fact that this is the very point where the meaning of "artificial intelligence" and "consumer behavior" are coupled in together. Due to the humanly nature of the users, when a decision is about be made everything comes down to them. The final users will define success. They will tell whether a system, a website, an application, an interface, a device, a service, a product, a customer support, a marketing campaign, an advertisement campaign, customer segmentation and user personalization sell or not. Hence, a closer look to user experience (UX) will help decision makers understand which entity in the entire AI dataflow is the most important and give extra attention to [57]. User experience (UX) is defined as the experience generated to users when they interact with systems, usually a graphical user interface, even if it a website, a web application, a software tool, or a robot. Despite the technological advantage the world has witnessed, the success of a company lies in the success of its systems interacting with the consumers. Consumers care mostly about what experience they get from the use of a system rather than caring about its complexity, sophistications or production and design cost. Customers' decision is affected by the efficacy of a system in its use and complexity. User experience is closely related to accessibility, usability, marketing, design, system performance, human machine interaction, utility, and human factors [57].

According to ISO.org, user experience is related to human centered designed quality factors. "Human-centered design aims to achieve required levels of human-centered quality. Human-centered quality is the collective term used to refer to usability, accessibility, user experience, and avoidance of harm from use" [58]. Another approach of UX is presented by the User Experience Professional Association. It internationally supports people who research, design, and evaluate the user experience (UX) of products and services [59].

Nevertheless, user experience is a valuable metric that can determine the success of a marketing plan. Customers want to follow the easiest steps to buy a product or service in the simplest way as possible having the best customer support they could possibly have. Artificial intelligence come to assist marketers personalize the purchasing procedures or lead the user to a certain path that will eventually bring conversions or trach users behavior in order to provide them real-time solution [36]. User personalization and segmentation provide determined contextual approach to specific users with conversion history and potential. AI evolves UX by patterning user behavior in a system e.g. a website or an app. The more a visitor stays in a website the merrier amount of data produces. Since machine learning models are constantly trained by data, this increase of data generation leads to better personalization and accurate predictions [36].

The user experience journey consists of interaction design, information architecture, marketing and communications, usability engineering, visual design, information design copywriting computer science (Fig. 10.4). AI-driven tools are able to gather this information from billions of users instantly. Customer experience will increase by the time each user will provide data to the computers. Algorithms will become "smarter", more accurate and customer satisfaction will continue to increase leading consequently to more purchases [39].

10.3.2 Consumer Behavior

The understanding of artificial intelligence contribution in consumer behavior lies in the explanation of the fundamental consumer behavior disciplines. Consumer behavior refers to human cognitive patterns and behavior engaging notions such as free will. It is believed that free will belongs to more generic forms of an abstract cultural social life, chronically expanded and individual's diversity roles. Free will encompasses self-control, carefully planned action, and logical decision [61]. Consumer behavior is defined as a complex and challenging field to be explained as preferences constantly vary. The traditional way of acquiring goods has been replaced with online selling which rapidly grows. The traditional marketing ways have been replaced by digital mediums and trends whereas the retailers and consumers meet online [61, 62].

Consumer behavior examines the factors that affect consumers' choice, under a free will spectrum, to purchase the product or service which satisfies their need. To be successful at any moment in a such a dynamic environment, the customers'

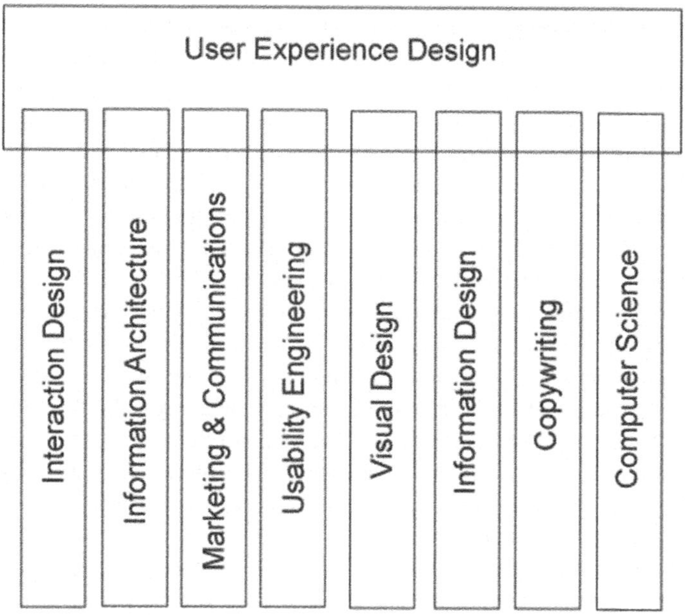

Fig. 10.4 The Boersma User Experience T-model. *Source* https://www.uxmatters.com/mt/archives/2011/10/the-t-model-and-strategies-for-hiring-ia-practitioners-part-1.php

journey must be pleasant, easy, and constant. In marketing science, the customers' journey, behavior, free will, decision making, and customer experience are bound inextricably together. This trend of product exchange has reduced the influence of middlemen over sales thereby gaining profit to the business and satisfaction to the consumers [62].

Marketing is an art of influence and it aims to find touch points where customers are open to influence. These touch points refer to primary potential set of choices that customers have, which known as the "funnel". The marketing goal it to reduce the primary funnel and lead the customers towards a narrowed set of choices of potential purchases as the consumers weigh the facts, consider the choices and finally purchase the product. However, due to well informed customers and the increase of digital channels, products, services, and globalization, the marketing landscape and the touch points are difficult to define [63, 64].

Nevertheless, the consumers' journey is much more complicated than the funnel suggests. Regarding the customer decision journey, consumers tend to get constantly bombarded by advertisements, brand campaigns, social media posts, news topics, comments, reviews, and recommendations. Referring to those who finally do not purchase, all this marketing effort ends up useless [63, 64].

However, there is a certain audience which finally completed a purchase. Due to certain key elements, customers are triggered to complete a transaction. Following the purchase, the post purchase phase determines the buyers' loyalty and the potential or re-engaging. Consumers' journey seems to be more of a repeated journey consisting of four stages: the initial decision process, the potential purchasing option evaluation, the conversion, and engagement [63, 64].

This is where the theory of messy middle emerges (Fig. 10.5). The term of "messy-middle" was introduced to show that the process that influences what customers will decide to buy is trivial and how consumers process the information is unclear. Currently, there is not enough information regarding purchase behavior. Marketers already know that the path between the triggering of a decision and conversion is not linear. Online touch points vary from buyer to buyer [65].

E-commerce grows along with the Internet and consumers can have information about almost everything. The way words are used in searches varies along with the actual meaning of these searched terms, so does the outcome. In an internet world where words can be interpreted in several meanings, the total exposure of a product or service is based on certain factors, different each time, yet trigger a potential purchase. This is where the messy-middle phase takes place. When and if the customers decide to make a purchase the experience from it could lead to customer loyalty and reengagement [65].

Consumers during their journey they seek for information that will help them decide what to purchase and they weight all the possible options. Messy middle

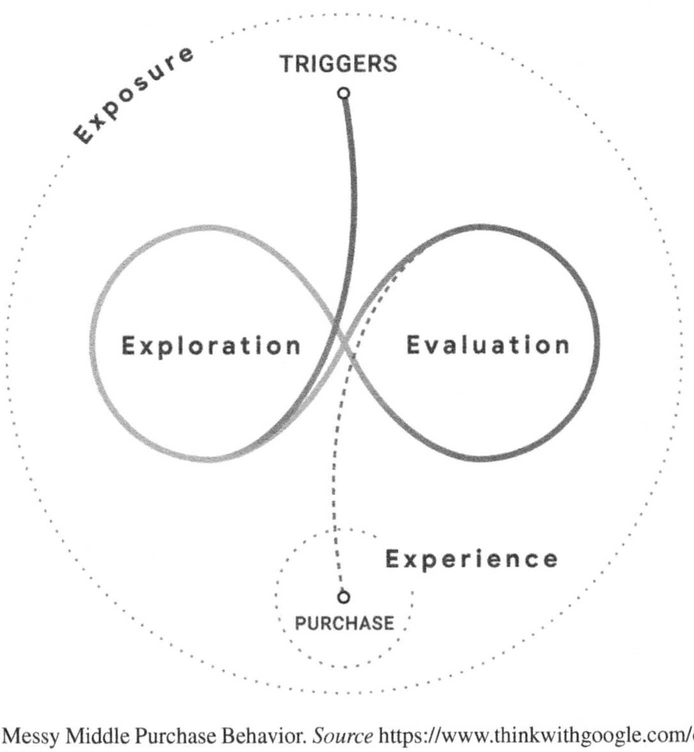

Fig. 10.5 Messy Middle Purchase Behavior. *Source* https://www.thinkwithgoogle.com/consumer-insights/navigating-purchase-behavior-and-decision-making

consists of exploration and evaluation (Fig. 10.5). Usually, people tend to search the web for information including websites, social media, technical reports, and search engines. These actions can be classified in one of the messy-middle repeated phases for as long as the consumers need to weight in the options and perform a decision [65].

The consumer decision journey has evolved into an extremely complexed procedure that has already set the foundations for a new overall marketing strategy adopted by the companies. Companies are currently realigned, trying to adjust to the new demanding circumstances. They have to embed new "smart" ways of consumers habits monitoring, brand effectiveness optimization and multiscale marketing transformations performance to be able to efficiently handle the amount of personalized data daily generated. AI application in consumer data will help brands analyze their customers, manage their funnel, optimize advertisement campaigns, increase users' satisfaction providing personalized solutions and product recommendations [63, 64].

10.3.3 AI in Consumer Behavior

AI applications on marketing science play a crucial role in conceiving the size of the impact that AI has on consumer behavior. Due to technological evolution consumers are able to search, find and select the best for them solutions to their needs. In this new era decision makers and researchers examine how they can increase user satisfaction. Huge volumes of web personalized data are constantly processed to reveal consumer behavior patterns that will allow buyers and customers experience qualitative experiences [79].

The consumers' journey examines and identifies customers' needs. AI enables decision makers to rapidly understand consumers current and future needs and target them through marketing campaigns. Websites, e-shops, web portals, social media and applications users' data contribute to user behavior and user intention identification, analysis and targeting. Through consumers targeting, AI manages to follow and observe consumers through certain stages of interaction during their decision journey. This allows AI to acquire data and make predictions through a set of stages. These stages include consideration, evaluation, purchase, and post purchase behavior. By the time marketers or advertiser realized that consumers have started their consuming journey, the consideration stage refers to consumers' reaching strategy using advertisements. Marketers use AI tools to specify the need and recommend solutions which are likely to be perceived by the audience as personalized recommendations that will satisfy their current need [8, 66]. Evaluation stage enables marketers to use AI tools for real-time image, voice, emotions, and natural language processing in order to generate content that will tip the balance and persuade customers that this is the best choice for them. Purchasing phase identifies what the consumers are ready to buy according to the season and adjust the price dynamically, along with brand authority, customer support and reassurances reinforcements. The final stage addresses consumers who assess the outcome of their decision to purchase. This assessment calculates the satisfaction from the purchasing experience, the satisfaction from product acquisition, the possibility of re-purchasing or even affecting others. Post purchasing also involves marketers' performance by assessing the level of effectiveness they provide in identifying and replying to the potential customers' needs [8].

10.3.3.1 AI Benefits in Consumer Behavior

Artificial intelligence revolutionizes how customers act and how companies operate. Artificial intelligence affects users experience and behavior. Organizations use AI to improve customer experience and increase customer satisfaction. The use of AI improves the performance includes certain benefits which affect the scale and the power of marketing methods assisting industries to accomplish their business goals. These benefits refer to sales increase, customer loyalty, customer convenience and customer trust. AI can identify consumers behavior, needs, feelings, decisions, and desires. By segmenting this information, it can target and provide personalized solutions specifically to a certain number or group of customers increasing the possibility of higher conversion rates. Customers segmentation, targeting and recommendations establish a "win–win" relationships between consumers and brands [67].

AI can increase customer loyalty to a point where both buyer and seller will enjoy a prosperous business relationship. Effective customer targeting combined with high quality product or service may increase customer loyalty in a way that the customers will continue reengaging with the brand enjoying its services. On the other hand, brand will expand their satisfied customers' funnel.

Since AI have been embedded in industries, the increasing users' convenience will endorse e-ecommerce by helping perform faster orders and purchases, suitable recommendations, and robotic shipping [67].

Nowadays, user privacy and data protection policy are the most important brands success factors. Both denotes trustworthiness for the brand. The way that brand store, protect, processes information, stay transparent, and protect customers data determines the level of success. When the product or service description respond as promised to the deliverable, the brand increase the possibility of building strong bonds with the consumers. AI contribute to that direction by offering services that clients find valuable and practical to their needs based on sophisticated data manipulation and information handling algorithms [68].

10.3.3.2 Factors Affect Online Consumer Behavior

The American Marketing Association presented "consumer behavior" as the "study of how customers, both individual and organizations, satisfy their needs and wants by choosing, purchasing, using and disposing of goods, ideas and services" [69].

When it comes to offline or online purchase, consumers behavior remains the same. But one some major differences are shopping environment and marketing communication. Consumers purchase decisions pass through certain stages including presale awareness, information retrieval, evaluation, decision, and post-sale behavior called engagement. Among these steps of there are certain factors that affect consumers decision to finally convert to customers [69].

As prementioned, customers journey is trivial and complex. Thus, there can be many reasons why consumers buy online. These have been summarized into four main categories including convenience, information, availability, and cost [70].

Convenience: Since e-shops operate 24 h a day, 30 days a month, 12 months a year, there is no time pressure to the consumers to search online for products or services. They are able to purchase anytime they want, getting the information they need and avoiding crowded places and long waiting queues. Due to personal reasons or psychological pressure a group of customers want to avoid close encounter with sales personnel [2, 70].

Information: Nowadays, customers can get easily informed. Due to the fact that customers do not have a closer view of the product, retailers provide much more information instead. Product reviews by other customers also play a game changer role to product or service acquisition [2, 70].

Availability: User interface, user experience, and purchasing procedure have become more human centric. Numerous products and services availability are also numerous helping consumers easily find what the search for [2, 70].

Cost: E-shops manage to provide their products cheaper than the physical stores basically because the number of employees providing support is smaller, the stores fixed costs are decreasing, and the competition increases as number of online store is constantly increasing [2, 70].

10.3.3.3 AI in Consumer Autonomy

AI evolution has brought new value to consumers and decision makers. AI affects consumers' behavior by automating processes making them consumer-centric and easier to use. Despite the advantages, AI due to data processing towards segmentation, targeting and recommendations is considered to be a threat to consumers' free will and autonomy. Consumers' autonomy in deciding for them and making their own choices seems to affect their wellbeing.

AI era involves three stakeholders: the consumers, the decision makers, and the algorithms. Consumers are able to find products or services to cover their need in low costs as never before. On the other hand, decision makers believe that since consumers segmentation lowers the search volumes and increase personalized recommendations it will genuinely increase transaction, and make decision making processes easier. AI algorithms collect, store, segment, analyze, pattern huge daily generated user datasets allowing decision makers make more balanced decisions [71, 79].

"Autonomous Customer 2015: On Hold for Intelligent Customer Service". Consumers consider the freedom of choice essential to their lives. Free will reinforces the peoples' belief that they make decisions proving their existence and independence. AI contribution to consumer autonomy must have the goal to serve the people.

Autonomous choices be balanced operating under certain rules and boundaries and most importantly endure self-determined choices [71, 79].

There are certain factors which affect consumers' autonomy to choose the best option for them through, a yet unexplained desire for choice, even if the selection procedure requires no effort. Consumers not only they feel a sense of hedonism during the purchasing procedure and the desire of choice, but they also significantly weight the autonomous choice as important [79].

Finally, seems that AI-driven personalized content and audience segmentation are on the customer's side assisting them to experience qualitative services *by responding to their needs for convenience, lower costs, product availability, better recommendations, and integrity* [71, 79].

10.3.4 AI and Ethics

In the era where artificial intelligence rapidly grows, governments and organizations have the responsibility to protect its citizens' lives and privacy by comprising a set of documents that include the ethical, constitution, legislation, regulations, and contracts which affect or are affected by AI. AI also impacts industry and customers, and trust must be built on strong foundations of data privacy and security, data transparency and safety. Additionally, the case of using biased data and "black box" problems are also big issues which need to be resolved [29].

Concerning ethics in business sector, companies make use of AI-driven applications and procedures in a way that will build trust between the brands and consumers. Data security requires companies to develop safety systems that will provide safely data storage and privacy protection for confidential personalized data to avoid leaking and unethical mistreatment [72].

Companies have also the responsibility to collect, store and protect the data under specific regulations and push governments towards that direction [73].

The most important issue refers to the lack of mechanism and data handling control. AI ethics guidelines lack law enforcement mechanisms and rules. Only ten out of 160 documents have binding agreements and enforcement mechanisms, where private and public sectors make voluntary commitments or provide documents with general recommendations (Fig. 10.6). Despite that the number of guidelines has been growing since 2019, the private sector relies heavily on voluntary commitments. Many state guidelines contain administrative recommendations presenting them as orientations aid or simple proposals weakening in this way the very meaning of the actual scope of the guidelines document [74].

Another serious issue that explains people's hesitation to accept AI is the fact that AI systems are built on machine-learning algorithms that are functionally unknown. Law control over AI "black box" will not manage to fully control the situation and hybrid solution is required. The introduction of big data makes privacy a difficult task. Due to their size and complexity big data cannot be easily controlled [75].

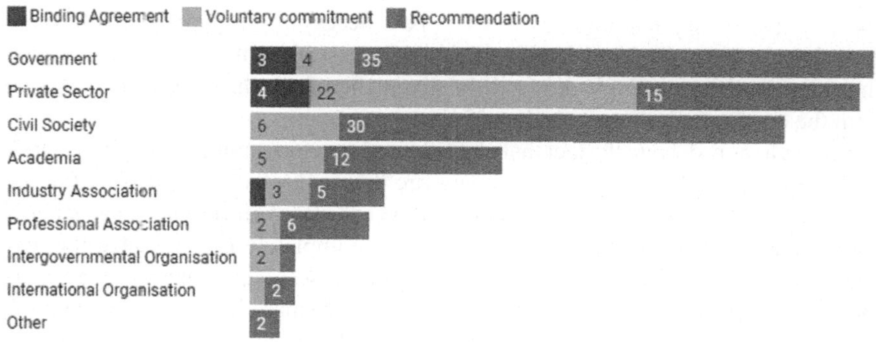

Fig. 10.6 AI Ethics Guidelines per Sector. *Source* https://algorithmwatch.org/en/ai-ethics-guidelines-inventory-upgrade-2020/

People expect companies to develop mechanisms to empower people in the marketplace by allowing them to have options of what data they will share and how these data will be processed or sold. Uncontrollable use of big data is a potential temptation for brands. Marketers would probably take advantage of this situation, leaving customers unprotected, by exploring their most sacred, deepest conscious or unconscious desires [75].

10.4 Conclusion

In a constantly changing world the emerging role of technology's disruptive nature is imperative to be highlighted. Since the world has changed the key to success lies on the sophisticated solutions that will be able to provide fast and accurate solutions. Humans and companies will be both benefited by this technological evolution. Purchase procedures and retail game changes. Businesses who will fail to adopt the new technological current will more than likely cease to be competitive [39].

AI driven applications assist customer and decisionmakers towards an amplified customers' journey full of opportunities, suggestions, and solutions. Personalized recommendations will save consumers' valuable time. Traditional retail will give its place to AI marketing and personalized commerce. The customers funnel increases day by day and the satisfied customers seem to find what they wanted; user friendly environments and interfaces, customer-centric tools and personalized suggestions and predictions through different mediums and channels. More than ever, daily interactions require the use of AI tools, otherwise the generated amount of data will be unmanageable [39].

Nevertheless, AI also transforms the way decision makers work and communicate allowing them to monitor, segment, and scale audiences to make better decisions and increase performance. AI has taken customer service to a whole new level using social media, email automation, visual and voice search, content generators and many other applications. AI application considered to be a necessity that brands should focus their future efforts. If not, brands will provide less relevant and time-consuming choices to customers. Customers will be dissatisfied from the experience acquired from the entire purchasing procedure [54].

In conclusion, despite the fact that algorithms outperform human capabilities, they are nothing more than a tool, and humans are those who have invented it after all. Artificial intelligence is on the rise, but it still needs more effort to become "smart". Nevertheless, it is expected AI will, eventually, completely revolutionize the way transactions take place. Brands, decision makers, researchers, and the consumers shape, willingly or not, the new AI landscape. Daily smartphone use, competition, research, and the urge of better solutions train AI algorithms [54].

The authors of this study hope and anticipate that this chapter will add a small "brick in the building" called "proper use of Artificial Intelligence towards the benefit of consumers' behavior" and motivate researchers, consumers, decision makers and policy makers towards a rational use of the technological means decision makers nowadays possess in order to preserve the right of human autonomy, dignity and independence.

References

1. V. Sankaran, Empowering consumer research with data science, in *Amity International Conference on Artificial Intelligence (AICAI)* (Dubai, United Arab Emirates, 2019), pp. 340–344, https://doi.org/10.1109/AICAI.2019.8701356
2. C. Katawetawaraks, L. Cheng, Online shopper behavior: influences of online shopping decision. Asian J. Bus. Res. **1** (2013). Available at SSRN: https://ssrn.com/abstract=2345198. Accessed 21 Aug 2020
3. S. Gochhait, O. Mazumdar, S. Chahal et al., Role of artificial intelligence (AI) in understanding the behavior pattern: a study on e-commerce, in *ICDSMLA 2019. Lecture Notes in Electrical Engineering*, vol. 601, eds. A. Kumar, M. Paprzycki, V. Gunjan (Springer, Singapore, 2020), https://doi.org/10.1007/978-981-15-1420-3_166
4. Rajanarthagi, Impact of artificial intelligence on digital marketing (2019), https://gecdesigns.com/blog/impact-of-artificial-intelligence-on-digital-marketing. Accessed 15 Aug 2020
5. L. Alexander, The Who, What, Why, & How of Digital Marketing (2020), https://blog.hubspot.com/marketing/what-is-digital-marketing. Accessed 21 Aug 2020
6. A. Orogun, O. Bukola, Predicting consumer behaviour in digital market: a machine learning approach. **8**:8391–8402 (2019). https://doi.org/10.15680/IJIRSET.2019.0808006
7. Oracle, Can virtual experiences replace reality? (2020), https://www.oracle.com/webfolder/s/delivery_production/docs/FY16h1/doc35/CXResearchVirtualExperiences.pdf?source=%3Aow%3Alp%3Acpo%3A%3A. Accessed 25 Aug 2020
8. J. Kietzmann, J. Paschen, E. Treen, Artificial intelligence in advertising: how marketers can Leverage artificial intelligence along the consumer journey. J. Advert. Res. **58**(3), 263–267 (2018). https://doi.org/10.2501/JAR-2018-035.J
9. S. Vázquez, O. Muñoz-García, I. Campanella et al., A classification of user-generated content into consumer decision journey stages. Neural Netw. **58**, 68–81 (2014), ISSN 0893-6080, https://doi.org/10.1016/j.neunet.2014.05.026
10. J.J. Sheu, Y.W. Chang, K.T. Chu, Applying decision tree data mining for online group buying consumers' behavior. Int. J. Electr. Cust. Relat. Manage. **2**(2), 140 (2008). https://doi.org/10.1504/ijecrm.2008.019929

11. R. Thiebaut, AI revolution: How data can identify and shape consumer behavior in ecommerce, (2019), https://doi.org/10.1108/978-1-78973-233-720191012
12. T. Taqwa, A. Suhendra, M. Hermita et al., Implementation of Naïve Bayes method for product purchasing decision using neural impulse actuator in neuromarketing, in *2015 International Conference on Information & Communication Technology and Systems (ICTS)* (Surabaya, 2015), pp. 113–118, https://doi.org/10.1109/ICTS.2015.7379882
13. R.A.E. Ahmeda, M.E. Shehaba, S. Morsya et al., Performance study of classification algorithms for consumer online shopping attitudes and behavior using data mining, in *2015 Fifth International Conference on Communication Systems and Network Technologies* (Gwalior, 2015), pp. 1344–1349, https://doi.org/10.1109/CSNT.2015.50
14. X. Niu, C. Li, X. Yu, Predictive analytics of e-commerce search behavior for conversion. AMCIS. H. Le-wei and L. Yu-sheng, Exploration in commercial websites on the association rules base on the consumers' behavior, in *2011 6th IEEE Joint International Information Technology and Artificial Intelligence Conference*, chongqing, 2011, (2017) pp. 42–44, https://doi.org/10.1109/ITAIC.2011.6030146
15. C. Hou, C. Chen, J. Wang, Tree-based feature transformation for purchase behavior prediction, in *IEICE Transactions on Information and Systems*, vol. E101.D (5), pp 1441–1444, Released, Online ISSN 1745-1361, Print ISSN 0916-8532, (2018), https://doi.org/10.1587/transinf.2017EDL8210
16. H. Valecha, A. Varma, I. Khare et al., Prediction of consumer behaviour using random forest algorithm, in *5th IEEE Uttar Pradesh Section International Conference on Electrical, Electronics and Computer Engineering (UPCON)* (Gorakhpur, 2018), pp. 1–6, https://doi.org/10.1109/UPCON.2018.8597070
17. P. Chandon, J.W. Hutchinson, E.T. Bradlow et al., Does in-store marketing work? Effects of the number and position of shelf facings on brand attention and evaluation at the point of purchase. J. Mark. **73**(6), 1–17 (2009). https://doi.org/10.1509/jmkg.73.6.1
18. K. Gidlöf, W. Annika, D. Richard et al., Using eye tracking to trace a cognitive process: gaze behaviour during decision making in a natural environment. J. Eye Mov. Res. **6** (2013)
19. S. Goyal, K.P. Miyapuram, U. Lahiri, Predicting consumer's behavior using eye tracking data, in *2015 Second International Conference on Soft Computing and Machine Intelligence (ISCMI)* (Hong Kong, 2015), pp. 126–129, https://doi.org/10.1109/ISCMI.2015.26
20. T. Kowatsch, W. Maass, In-store consumer behavior: how mobile recommendation agents influence usage intentions, product purchases, and store preferences. Comput. Hum. Behav. **26**(4), 697–704 (2010), ISSN 0747-5632, https://doi.org/10.1016/j.chb.2010.01.006
21. D. Liciotti, E. Frontoni, A. Mancini et al., Pervasive system for consumer behaviour analysis in retail environments, in *Video Analytics. Face and Facial Expression Recognition and Audience Measurement. VAAM 2016, FFER 2016*. Lecture Notes in Computer Science, ed. K. Nasrollahi et al. vol. 10165 (Springer, Cham, 2017), https://doi.org/10.1007/978-3-319-56687-0_2
22. Ł. Łysik, R. Kutera, P. Machura, Behavioural and technical factors of influence on purchase behaviour of mobile consumers, in *Proceedings of the 18th International Academic Mindtrek Conference on Media Business, Management, Content & Services* (AcademicMindTrek '14, 2014), https://doi.org/10.1145/2676467.2676494
23. L. Muzellec, E. O'Raghallaigh, Mobile technology and its impact on the consumer decision-making journey. J. Advert. Res. **58**(1), 12–15 (2017). https://doi.org/10.2501/jar-2017-058
24. M. Zhang, G. Chen, Q. Wei, Discovering consumers' purchase intentions based on mobile search behaviors, in *Flexible Query Answering Systems 2015. Advances in Intelligent Systems and Computing*, ed. T. Andreasen et al., vol. 400 (Springer, Cham., 2016) https://doi.org/10.1007/978-3-319-26154-6_2
25. S. Moro et al., Predicting social media performance metrics and evaluation of the impact on brand building: a datamining approach. J. Bus. Res. (2016). https://doi.org/10.1016/j.jbusres.2016.02.010
26. C. Kim, S.U. Yang, Like, comment, and share on Facebook: how each behavior differs from the other. Public Relations Rev. **43**(2), 441–449 (2017). https://doi.org/10.1016/j.pubrev.2017.02.006

27. K. Hao, What is AI? We Drew You a Flowchart to Work it Out (2018), https://www.technologyreview.com/2018/11/10/139137/is-this-ai-we-drew-you-a-flowchart-to-work-it-out. Accessed 26 Aug 2020
28. B.J. Copeland, Artificial Intelligence (2020), https://www.britannica.com/technology/artificial-intelligence. Accessed 15 Aug 2020
29. Accenture, Artificial Intelligence (2020), https://www.accenture.com/us-en/insights/artificial-intelligence-summary-index. Accessed 15 Aug 2020
30. Nexus Integra, Advantages and disadvantages of artificial intelligence (2019), https://nexusintegra.io/advantages-disadvantages-artificial-intelligence. Accessed 15 Aug 2020
31. H. Jasim, H. Hiba, H. Shnain et al., (2015) Big data and 5 V's characteristics. 2393–2835
32. J. Hall, How Artificial Intelligence is Transforming Digital Marketing (2019), https://www.forbes.com/sites/forbesagencycouncil/2019/08/21/how-artificial-intelligence-is-transforming-digital-marketing/#26a4a36421e1. Accessed 13 Aug 2020
33. The Keen Folks, AI Marketing: How to Leverage New Technology to Maximize Results (2020), https://thekeenfolks.com/keensights/artificial-intelligence-marketing. Accessed 15 Aug 2020
34. Smartboost, How to Predict Consumer Behavior Using AI Marketing (2020), https://smartboost.com/blog/how-to-predict-consumer-behavior-using-ai-marketing. Accessed 15 Aug 2020
35. K. Baker, The Ultimate Guide to Content Marketing in 2020 (2020), https://blog.hubspot.com/marketing/content-marketing. Accessed 15 Aug 2020
36. C. Myers (2019) How Will a Transform the Digital Marketing in 2019? https://www.digitaldoughnut.com/articles/2019/february/how-will-ai-transform-the-digital-marketing. Accessed 15 Aug 2020
37. J. Miller, 5 Ways AI is Transforming Marketing (2018), https://aibusiness.com/document.asp?doc_id=760620&site=aibusiness. Accessed 15 Aug 2020
38. Google Analytics, Data Import (2020), https://support.google.com/analytics/topic/6065609?hl=en&ref_topic=1727148&authuser=0. Accessed 15 Aug 2020
39. M. Olmez, How to Use AI in E-Commerce (2018), https://www.accenture.com/nl-en/blogs/insights/how-to-use-ai-in-e-commerce. Accessed 15 Aug 2020
40. E. Van den Broeck, K. Poels, M. Walrave, A factorial survey study on the influence of advertising place and the use of personal data on user acceptance of Facebook ads. Am. Behav. Sci. **61**, 000276421771756 (2017). https://doi.org/10.1177/0002764217717560
41. Smart Insights, Chatbot Conversations Just Got More Human (2020), https://www.smartinsights.com/user-experience/chatbot-conversations-just-got-human. Accessed 24 Aug 2020
42. B. Zarouali, E. Van den Broeck, M. Walrave et al., Predicting consumer responses to a chatbot on Facebook. Cyberpsychol. Behav. Soc. Network. **21** (2018) https://doi.org/10.1089/cyber.2017.0518
43. OMDIA TRACTICA, Artificial Intelligence Software Revenue to Reach $59.8 Billion Worldwide by 2025 (2017), https://tractica.omdia.com/newsroom/press-releases/artificial-intelligence-software-revenue-to-reach-59-8-billion-worldwide-by-2025. Accessed 15 Aug 2020
44. Juniper Research, Chatbot Conversations to Deliver $8 Billion in Cost Savings by 2022 (2020), https://www.juniperresearch.com/new-trending/analystxpress/july-2017/chatbot-conversations-to-deliver-8bn-cost-saving. Accessed 18 Aug 2020
45. A. Butt, Artificial Intelligence, Natural Language Processing, Machine Learning and Voice Search. A Match Made in Heavens (2019), https://quoracreative.com/article/voice-search-evolution. Accessed 15 Aug 2020
46. S. Mohanadasan, Visual Search Guide: Who Uses it, Benefits, and Optimization Tip (2020), https://www.semrush.com/blog/visual-search-guide-benefits-optimizatiion/#header2. Accessed 22 Aug 2020
47. L. Heller, Visual Search Wins Over Text as Consumers' Most Trusted Information Source (2019), https://www.businesswire.com/news/home/20190204005613/en/Visual-Search-Wins-Text-Consumers%E2%80%99-Trusted-Information. Accessed 15 Aug 2020
48. C. Boyd, Visual Search - The Ultimate Guide: Statistics, News, Trends, and Tips (2018), https://medium.com/@clarkboyd/visual-search-trends-statistics-tips-and-uses-in-everyday-life-d20084dc4b0a. Accessed 15 Aug 2020

49. M. Kaput, Artificial Intelligence in Email Marketing (2021). https://www.marketingaiinstitute.com/blog/ai-in-email-marketing. Accessed 27 Jan 2021
50. S.R. Chowdhuri, The Impact of Artificial Intelligence on Social Media (2020), https://www.manipalprolearn.com/blog/impact-artificial-intelligence-social-media. Accessed 11 Aug 2020
51. PPCExpo, The Ultimate Guide to Google Smart Campaigns (2020), https://ppcexpo.com/blog/smart-campaigns. Accessed 15 Aug 2020
52. Facebook for Business, Making Creative Impact on Facebook Platforms with Text, Timing and Format (2020), https://www.facebook.com/business/news/insights/making-creative-impact-on-facebook-platforms-with-text-timing-and-format. Accessed 15 Aug 2020
53. Google Ads, Smart Campaigns (2020), https://support.google.com/google-ads/answer/9367181?hl=en&ref_topic=9024774. Accessed 15 Aug 2020
54. K. Novoseltseva, Artificial Intelligence in E-Commerce: Benefits, Statistics, Facts, Use Cases & Case Studies (2018), https://apiumhub.com/tech-blog-barcelona/artificial-intelligence-ecommerce. Accessed 15 Aug 2020
55. PAT Research, What is predictive analytics (2020), https://www.predictiveanalyticstoday.com/what-is-predictive-analytics. Accessed 25 Aug 2020
56. C. Crawford, Social Media Analytics: The Complete Guide (2020), https://www.socialbakers.com/blog/social-media-analytics-the-complete-guide. Accessed 21 Aug 2020
57. J. Gube, What is User Experience Design? Overview, Tools and Resources (2010), https://www.smashingmagazine.com/2010/10/what-is-user-experience-design-overview-tools-and-resources. Accessed 21 Aug 2020
58. ISO, ISO 9241-220:2019 Ergonomics of Human-System Interaction — Part 220: Processes for Enabling, Executing and Assessing Human-Centred Design Within Organizations (2020), https://www.iso.org/obp/ui/#iso:std:iso:9241:-220:ed-1:v1:en:sec:E. Accessed 16 Aug 2020
59. UXPA, User Experience Professionals Association (2020), http://uxpa.org. Accessed 15 Aug 2020
60. N. Davis, The T-Model and strategies for hiring a practitioners: Part 1 (2011), https://www.uxmatters.com/mt/archives/2011/10/the-t-model-and-strategies-for-hiring-ia-practitioners-part-1.php. Accessed 21 Aug 2020
61. R.F. Baumeister, E.A. Sparks, T.F. Stillman et al., Free will in consumer behavior: Self-control, ego depletion, and choice. J. Consum. Psychol. **18**, 4–13 (2008). https://doi.org/10.1016/j.jcps.2007.10.002
62. P. Mani, Consumer behaviour. Int. J. Innova. Res. Comput. Commun. Eng. **8** 4 (2019), https://doi.org/10.35940/ijitee.L1127.10812S19
63. M. Brenner, Top Benefits & Uses of AI in Digital Marketing (2020), https://marketinginsidergroup.com/artificial-intelligence/5-benefits-of-ai-for-digital-marketers/. Accessed 15 Dec 2020
64. D. Court, D. Elzinga, S. Mulder, O.J. Vetvik, The Consumer Decision Journey (2009), https://www.mckinsey.com/business-functions/marketing-and-sales/our-insights/the-consumer-decision-journey. Accessed 19 Aug 2020
65. A. Rennie, J. Protheroe, How people decide what to buy lies in the 'messy middle' of the purchase journey (2020), https://www.thinkwithgoogle.com/consumer-insights/navigating-purchase-behavior-and-decision-making. Accessed 22 Aug 2020
66. P. Hudson, The Impact of Artificial Intelligence on Consumer Behavior (2020), https://blog.flexmr.net/ai-consumer-behaviour. Accessed 15 Jan 2021
67. Robotic Marketer, Does artificial intelligence influence consumer behavior? (2020), https://www.roboticmarketer.com/does-artificial-intelligence-influence-consumer-behavior. Accessed 21 Aug 2020
68. K. Szatylowicz, 5 ways retail AI Will Influence Consumer Behavior (2019), https://blog.crobox.com/article/ai-consumer-behavior. Accessed 15 Aug 2020
69. AMA, Consumer Behavior (2020), https://www.ama.org/topics/consumer-behavior. Accessed 15 Aug 2020
70. G. Asamoah, George Asamoah Factors Which Influence the Buying Behaviours of Customers with Multiple Regular Customer Cards (2012), https://core.ac.uk/download/pdf/38067117.pdf. Accessed 15 Jan 2021

71. C. Everett, What is the 'Autonomous Customer' - and Why is it Important? (2015), https://www.mycustomer.com/marketing/strategy/what-is-the-autonomous-customer-and-why-is-it-important. Accessed 15 Aug 2020
72. Data Ethics, Danish Companies Behind Seal for Digital Responsibility (2019), https://dataethics.eu/danish-companies-behind-seal-for-digital-responsibility. Accessed 15 Aug 2020
73. M. Alemany Oliver, J. Vayre, Big data and the future of knowledge production in marketing research: ethics, digital traces, and abductive reasoning. J. Market Anal. **3**, 5–13 (2015). https://doi.org/10.1057/jma.2015.1
74. Algorithm Watch, In the Realm of Paper Tigers – Exploring the Failings of AI Ethics Guidelines (2021), https://algorithmwatch.org/en/ai-ethics-guidelines-inventory-upgrade-2020. Accessed 21 Jan 2021
75. Y. Bathaee, The artificial intelligence black box and the failure of intent and causation. Harvard J. Law Technol. **31**, 889 (2018)
76. M. Kaput, AI for Advertising: What You Need to Know (2020), https://www.marketingaiinstitute.com/blog/ai-in-advertising. Accessed 17 Aug 2020
77. M. Kaput, AI for Social Media: What You Need to Know (2020), https://www.marketingaiinstitute.com/blog/ai-for-social-media. Accessed 13 Aug 2020
78. S. Okazaki, A.M. Díaz-Martín, M. Rozano et al., Using Twitter to engage with customers: a data mining approach. Internet Res. **25**(3), 416–434 (2015). https://doi.org/10.1108/intr-11-2013-0249
79. Q. André, Z. Carmon, K. Wertenbroch et al., Consumer choice and autonomy in the age of artificial intelligence and big data. Cust. Needs Solut. (2017). https://doi.org/10.1007/s40547-017-0085-8
80. L.B. Said, T. Bouron, A. Drogoul, Agent-based interaction analysis of consumer behavior, in *Proceedings of the First International Joint Conference on Autonomous Agents and Multiagent Systems Part 1 - AAMAS '02* (2002). https://doi.org/10.1145/544741.544787
81. Smart Insights, How Artificial Intelligence is Reshaping Email Marketing (2019), https://www.smartinsights.com/email-marketing/how-artificial-intelligence-is-reshaping-email-marketing. Accessed 25 Aug 2020
82. Smart Insights, Key Features Putting E-Commerce Personalization at the Top of the Retail Agenda (2019), https://www.smartinsights.com/ecommerce/key-features-putting-e-commerce-personalization-at-the-top-of-the-retail-agenda. Accessed 23 Aug 2020
83. Smart Insights, Personalization is Key to Improving Lead Generation Quality for 60% of Marketers (2019), https://www.smartinsights.com/lead-generation/lead-generation-strategy/personalization-key-improving-lead-generation. Accessed 28 Aug 2020
84. Statista, Big Data Revenue Worldwide from 2016 to 2027, by Major Segment (2020), https://www.statista.com/statistics/301566/big-data-factory-revenue-by-type. Accessed 20 Aug 2020
85. Statista, Volume of Data/Information Created Worldwide from 2010 to 2024 (2020), https://www.statista.com/statistics/871513/worldwide-data-created. Accessed 29 Aug 2020

Part III
Theoretical Advances in Computation and System Modeling

Chapter 11
Coupled Oscillator Networks for von Neumann and Non-von Neumann Computing

Michele Bonnin, Fabio Lorenzo Traversa, and Fabrizio Bonani

Abstract The frenetic growth of the need for computation performance and efficiency, along with the intrinsic limitations of the current main solutions, is pushing the scientific community towards unconventional, and sometimes even exotic, alternatives to the standard computing architectures. In this work we provide a panorama of the most relevant alternatives, both according and not to the von Neumann architecture, highlighting which of the classical challenges, such as energy efficiency and/or computational complexity, they are trying to tackle. We focus on the alternatives based on networks of weakly coupled oscillators. This unconventional approach, already introduced by Goto and von Neumann in the 1950s, is recently regaining interest with potential applications to both von Neumann and non-von Neumann type of computing. In this contribution, we present a general framework based on the phase equation derived from the description of nonlinear weakly coupled oscillators, especially useful for computing applications. We then use this formalism to design and prove the working principle and stability assessment of Boolean gates such as NOT and MAJORITY, that can be potentially employed as building blocks for both von Neumann and non-von Neumann architectures.

M. Bonnin (✉) · F. Bonani
Department of Electronics and Telecommunication, Politecnico di Torino,
Corso Duca degli Abruzzi 24, 10129 Turin, Italy
e-mail: michele.bonnin@polito.it

F. Bonani
e-mail: fabrizio.bonani@polito.it

F. L. Traversa
MemComputing Inc, 9909 Huennekens Street, Suite 110, 92121 San Diego, CA, USA
e-mail: ftraversa@memcpu.com

M. Virvou et al. (eds.), *Advances in Artificial Intelligence-based Technologies*,
Learning and Analytics in Intelligent Systems 22,
https://doi.org/10.1007/978-3-030-80571-5_11

11.1 Introduction

Electronic architectures designed to perform specialized or general computational tasks date back to the end of the first half of the XX century, when the University of Pennsylvania disclosed ENIAC [1, 2]. However, it is only few months after the ENIAC presentation, during the session of lectures at UPenn titled "The Theory and Techniques for Design of Digital Computers", that von Neumann wrote the *First Draft of a Report on the EDVAC* [3] starting what we can dub the von Neumann architecture age.

For almost a century, the von Neumann architecture has been the standard reference for the design of electronic computing machines, especially for the general purpose ones. Its most basic description encompasses an input module, a central processing unit, a memory bank and an output module. Sets of instructions (programs) can be written in the memory, and the processing unit accesses the program and processes input data performing a sequence of operations, including reading and writing repeatedly the memory, before ending the program and returning the output [4].

The von Neumann architecture can be also viewed as the ideal hardware design of a Turing machine [4, 5], and as such it inherits the related versatility along with the limitations. In fact, there is no known design that can be used to create a better artificial general purpose computing machine and, at the same time, be completely and deterministically controllable and programmable. However, the same design naturally reveals its drawbacks, such as the famous *von Neumann bottleneck* [4, 6], dictating that the system throughput is limited by the data transfer between CPU and memory, as the majority of the computation energy is used for the data movement rather than for the actual computation [4, 7, 8]. Limitations do not come only from physics, but also from the computational scaling capability of these machines. For example, the von Neumann architectures are sequential in essence (notice that CPUs exploiting vectorization or distributed architectures like GPUs are still sequential in nature because, even if they can process a set of elements in parallel, the set size does not scale with the problem size) and they work with a finite set of symbols to encode information. On the one hand, both these features allow von Neumann architectures to handle virtually any decidable problem size (problems are termed decidable if they can be computed by a Turing machine [5]). On the other hand, many of these problems become quickly intractable for von Neumann architectures as their size grows. The theory of computational complexity based on Turing machines [5], allows us to categorize decidable problems into classes of different polynomial scaling as a function of the problem size. Therefore, there are problems that can become intractable with von Neumann architectures either because the polynomial degree is too large for the typical problem size, or, even worse, if the scaling rule is a "non-deterministic" polynomial, as for the NP problem classes [5], thus making them sometimes intractable even for very small size. For instance, solving a square system of linear equations in n variables is a problem of complexity $O(n^3)$, i.e. its solution with a von Neumann architecture needs an order n^3 operations. Therefore, if $n = 10^6$ we would need to perform an order 10^{18} operations: this would require a time $O(10^9)$ s using a CPU working at GHz clock, i.e. tens of years!

This picture highlights some among the most important reasons behind the quest of new solutions for computing devices, architectures and paradigms. The exponential growth of computation demand is everywhere, from industrial to consumer electronics, from cloud to edge, the internet of things, autonomous vehicles and much more. All of them require a day by day performance increase, along with energy efficiency and reliability. The demand is already challenging the current computing systems, the Moore's law era is at the end and projections for the near future clearly show that either new solutions will emerge, or the demand increase will soon become unsustainable [4, 7, 8].

Recent years have seen many, novel or rediscovered, alternative solutions proposed by academia or industry giving some hope for satisfying near future computation demands. However, no clear winner is there yet. On the contrary, it looks like many specialized hardware and architectures are emerging, and there will be multiple winners depending on the applications or problems to be tackled. This is not surprising, since it is quite obvious that a price is paid for a machine to be general purpose.

GPU is a nice example of a specialized computing architecture already introduced in the 1970s Initially it was designed to accelerate specific graphic tasks by means of a distributed circuit design characterized by a very large memory bandwidth and a specialization in performing the algebraic operations needed for graphics, such as matrix-vector multiplication [9]. In the last decade, GPUs are increasingly used for other type of applications, such as simulation and machine learning: industry and academia are investing a great deal of effort to develop more sophisticated, efficient and powerful GPUs, so that today they represent the most advanced form of high performance computing [9, 10].

The hype on artificial intelligence, recently invigorated by the renewed development of deep learning and artificial neural networks [11, 12], is driving industry and academia towards new computing architectures specialized solely for the training or the inference of neural networks [13, 14]. For these applications two main approaches exist: digital, such as Google's TPU [15, 16], and analogue, as for instance the memristor crossbar neural networks on chip [17–19].

An example of a revived alternative, suggested in the 1980s by Benioff, Feynman and Manin [20–22], is quantum computing. The idea is to leverage quantum mechanics to create a non-Turing machine that redefines the computational complexity classes for some problems [23]. For example, in 1994 Shor showed that, using entangled qubits and quantum gates for their manipulation, it is possible to set up a quantum circuit capable of factorizing integer numbers in cubic time with respect to the length (in digits) of the integer number [24]. Obviously, this was a great theoretical achievement, since it proves that a quantum circuit could break RSA encryption [5, 24]. However, although quantum circuits seem a very appealing alternative, the technological barrier to their development has proved impervious: only in the last decade, few universities and organizations have accepted the challenge and are trying to build prototypes. Nevertheless, it is still in its infancy, and it will probably require decades before commercial-grade quantum computers will be available [25].

It is also worth mentioning the recent effort in neuromorphic computing, where a central role is played by asynchronous digital or analogue circuits trying to mimic brain neural networks [26, 27]. Noticeable examples are systems like SpiNNaker developed at the university of Manchester [28], or Loihi by Intel Labs [29], or memristive systems at IBM [30] that implement spiking neural networks [31].

A general path aimed at mitigating the von Neumann bottleneck goes under the name of near-memory computing [32], with the limit case of in-memory computing [33]. The idea here is to bring as close as possible memory and processing unit in order to reduce the energy and latency due to the data exchange between them [32]. For in-memory computing, at least some logic is directly performed in memory [33–36].

A step forward beyond in-memory computing is the emerging technology dubbed memcomputing. It was introduced as a non-Turing paradigm [37], and it represents the most idealized model of a computational memory [37–39]. It can be realized in analog [40] or digital form [41], however its most effective realization is through non-von Neumann architectures called self-organizing gates [41–43]. The working principle of these gates is as follows: they are designed to reach an equilibrium such that the terminal states are consistent with some prescribed relation [41, 42]. For example, self-organizing *logic* gates reach an equilibrium when the terminals satisfy a prescribed set of Boolean relations [41]. However they are terminal agnostic, in the sense that they accept a superposition of input and output signals at each terminal [41, 42]. This allows to assemble self-organizing circuits by interconnecting self-organizing gates [38, 41, 42] with controllable properties of the dynamics [44, 45]. This novel computing architecture has proven to be particularly efficient in solving combinatorial optimization problems [46–50].

Almost all the emerging approaches so far mentioned have in common the exploitation of a transistor used as fundamental building block to create gates or other basic computing structures that ultimately reach some static working point. Alternatives to this scheme are for example the p-bits introduced by Datta *et al.*, [51], that can be viewed as a stochastic alternative realization of memcomputing through CMOS-assisted nanomagnets [51] or magnetic tunnel junctions [52]. A similar concept has been developed at Los Alamos National lab using artificial spin ice systems [53].

This quest for alternative, non-conventional computing solutions has also recently revived the use of oscillators as building blocks for both von Neumann and non-von Neumann architectures. They were initially introduced independently by Goto [54, 55] and von Neumann [56, 57] in the 1950s. The idea in this case is to encode information in the relative phase among different oscillators and, by means of proper coupling, use them to perform basic operations such as the NOT or MAJORITY functions [54–57]. In the 1960s, machines called Parametrons implementing the Goto design were built in Japan [54, 55]. Parametrons saw some successes, however they were soon eclipsed by digital computers exploiting the first generation of integrated semiconductor devices. Today, the interest is renewed not only because there are many other modern and more compact ways to integrate oscillators, ranging from ring oscillators [58] to spin-torque [59] and laser-based [60] structures and beyond

[61–66], but also because coupled oscillators potentially represent a very low power computing system when employed in von Neumann architectures [67–69].

However, interest in coupled oscillators is significant not only for von Neumann architectures, as recently they have been listed among the most promising building blocks for non-von Neumann architectures such as Ising machines [70–72] and bio-inspired computing architectures [73–76].

In this work we focus on the general mathematical framework that is used to describe the dynamics of networks of oscillators employed to perform computation in the von Neumann as well as non-von Neumann sense. We introduce general concepts such as self-sustained oscillations, limit cycles and stability to describe a single autonomous system, and then we extend these to (weakly) coupled oscillatory systems in Sect. 11.2 where we also describe examples of network configurations, e.g. master-slave, useful for certain types of computations. In Sect. 11.3 we introduce the phase equation through the concept of the isochron, and we show how to write a general phase-amplitude equation system that describes the dynamics of a network of weakly coupled oscillators. In Sect. 11.4 we discuss how to design a simple network of oscillators to perform Boolean logic by means of a register and a complete base formed of the NOT and the MAJORITY gates. Moreover, we also show how to obtain stable OR and AND gates from the MAJORITY operation. For each case, we discuss the working principle showing the asymptotic stability properties of the phase equations describing the network. We finally draw our conclusions in Sect. 11.5.

11.2 Basic Unit, Network Architecture and Computational Principle

Oscillators are electronic circuits able to produce a continuous, repeated, alternating waveform without any input. Oscillators periodically convert energy between two different forms, and the energy dissipated during each oscillation cycle can, and should, be significantly less than the total energy involved in the oscillation. For example, an LC oscillator characterized by a quality factor Q loses approximately a fraction equal to $1/Q$ of its energy in each oscillation cycle.

Every electronic oscillator requires at least two energy-storage elements and at least one *nonlinear element*. Fig. 11.1a shows the basic computing unit that we shall consider. Since both inductor and capacitor are linear and passive (i.e., $L > 0$ and $C > 0$), the resistive one-port N_R must be active, i.e., its characteristic should be contained at least partially in the second and/or fourth quadrant of the (v, i) plane for oscillation to be possible [77]. Besides, the characteristic curve linking v and i is required to be nonlinear. An example is shown in Fig. 11.1b: As the current through the resistor increases, the forward voltage increases rapidly until it reaches a maximum. As the current is further increased, the voltage starts decreasing, until it reaches a minimum, and then it starts increasing again. The region between the peak and the minimum is characterized by a negative differential resistance, and thus the resulting circuit is dubbed *negative resistance oscillator.*

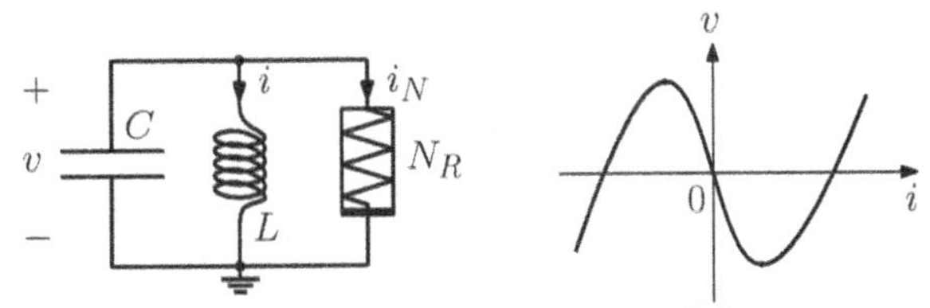

Fig. 11.1 **a** Basic unit: *LC* oscillator along with a nonlinear resistor. **b** Nonlinear $v(i)$ characteristic with a negative differential resistance region for N_R

The circuit exhibits *self-sustained oscillations* if the energy dissipated during the time in which it operates in the positive differential resistance regions, equals the energy supplied by N_R when it operates in the negative differential resistance region. Self-sustained oscillations correspond to a *limit cycle,* i.e. an isolated periodic orbit in the oscillator state space. Proving the existence of limit cycles is, in general, a very difficult, if not impossible, task. For the circuit in Fig. 11.1, it is however possible to find conditions for the existence of limit cycles. Applying Kirchhoff current law to one of the two nodes we obtain

$$C\frac{\mathrm{d}v}{\mathrm{d}t} + i_N + i = 0 \tag{11.1}$$

Deriving again with respect to time and using the nonlinear resistor characteristic $i_N = g(v)$,

$$\frac{\mathrm{d}^2 v}{\mathrm{d}t^2} + \frac{1}{C}\frac{\mathrm{d}g(v)}{\mathrm{d}v}\frac{\mathrm{d}v}{\mathrm{d}t} + \frac{1}{LC}v = 0 \tag{11.2}$$

Assuming a cubic nonlinearity $g(v) = -g_1\, v + g_3\, v^3$, where g_1 and g_3 are positive parameters, (11.2) is a *Liénard equation*, which is known to admit a stable limit cycle surrounding the origin [78].

For many practical purposes, it is more convenient to rewrite the second order, ordinary differential equation (11.2), as a system of two, first order differential equations

$$\frac{\mathrm{d}i}{\mathrm{d}t} = \frac{1}{L}v \tag{11.3a}$$

$$\frac{\mathrm{d}v}{\mathrm{d}t} = -\frac{1}{C}i + \frac{1}{C}g_1\, v - \frac{1}{C}g_3 v^3 \tag{11.3b}$$

Introducing the scaled variables

$$x = i, \qquad y = \sqrt{\frac{C}{L}}v \qquad \tau = \frac{1}{\sqrt{LC}}t \tag{11.4}$$

system (11.3) becomes

$$\frac{dx}{d\tau} = y \tag{11.5a}$$

$$\frac{dy}{d\tau} = -x + G(y) = -x + G_1 y - G_3 y^3 \tag{11.5b}$$

where $G_1 = g_1\sqrt{L/C}$ and $G_3 = g_3(L/C)^{3/2}$.

For a wide range of values of parameters G_1 and G_3, the dynamical system (11.5) not only has a stable limit cycle, but it is also *structurally stable,* meaning that the qualitative behavior of the trajectories is unaffected by small perturbations of the vector field. In particular the amplitude (but not the period) of the limit cycle is robust to external perturbations, showing small variations when perturbation terms are added to the vector field. This is a key requirement for computation purposes.

To implement computation, oscillators must exchange information, i.e. they must be coupled to form a network. Connections between units must be weak, meaning that they should exchange small energy amounts, to avoid possible bifurcations of limit cycles. One basic assumption is that coupling is so weak that it affects only the phases of oscillators, without modifying significantly amplitude and frequency. A second major assumption to be made concerns how the oscillators are connected, that is, the network topology. The architecture plays a fundamental role in determining the network dynamics, and investigating the role played by interaction strength and topology is a rich and challenging research area (see [79, 80] and references therein). On the one hand, the possibility to choose coupling schemes, interaction range and strength, gives the designer a wide range of possibilities to devise novel computational schemes. On the other hand, physical constraints must be taken into account. In a highly interconnected structure wirings may become burdensome, because the number of interconnects outweighs that of the neurons. Moreover, the mathematical modelling and analysis complexity grows exponentially with the number of neurons and connections, thus making the problem quickly intractable.

We shall consider networks with both local and global connections. In particular, the network has a central unit that will be used as a *reference*, connected to all other *neurons* as depicted in Fig. 11.2. A crucial point is that connections between the reference oscillator and neurons are one-directional, that is, the reference unit influences all other nodes, but not vice-versa. This kind of connection is also referred to as *master-slave*. The resulting graph is *directed* and *weighted.* Neurons have local connections with their first neighbors only. Such an architecture is inspired to biological neural structures, e.g. the thalamus-cortex system, and has been previously suggested as a promising topology for unconventional computing solutions [81, 82].

Information is encoded into the phase of each oscillator. Because the phase is a continuous variable, oscillatory networks in general operate as analog, or in modern terminology *bio-inspired*, computers. However, by a proper engineering of the couplings, it is possible to design oscillatory networks where only discrete values of the phase differences are attainable, i.e. those corresponding to asymptotically stable states, obtaining oscillatory networks that operate as digital computers based on Boolean logic.

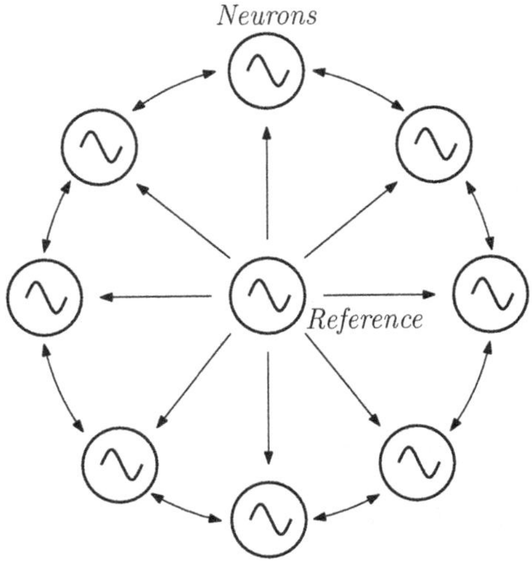

Fig. 11.2 Network architecture

Information is processed modifying the phase relationships among oscillators through the action of couplings. It should be noted that the simple presence of a physical connection between two oscillators does not guarantee by itself that they influence each other. Interaction occurs if the coupling strength exceeds a threshold that depends on the frequency mismatch. In other words, two oscillators whose free running frequencies are sufficiently apart, may be transparent one to the other, even if they are physically connected. Therefore it is possible, in principle, to control couplings without acting on the physical connections, but rather by simply modifying the free running frequencies of the individual oscillators.

11.3 Nonlinear Oscillator Networks and Phase Equation

The most important concept to be defined for implementing computation with coupled oscillators is the phase concept. Defining the phase of a nonlinear oscillator is a nontrivial problem, whose analysis dates back to the seminal works [66, 83]. Today, there is a large consensus that the most convenient way to define the phase of nonlinear oscillators is based on the concept of *isochron* [84–93].

Consider a nonlinear oscillator with an asymptotically stable, T periodic limit cycle $\boldsymbol{x}_s$. Take a reference initial condition $\boldsymbol{x}_s(0)$ on the limit cycle, and assign phase zero to this point, i.e. $\phi(\boldsymbol{x}_s(0)) = 0$. The state of the oscillator with initial condition $\boldsymbol{x}_s(0)$ at time t is $\boldsymbol{x}_s(t)$, with phase $\phi(\boldsymbol{x}_s(t)) = 2\pi t/T = \omega t$.

Definition 1 The *isochron* based at $\boldsymbol{x}_s(0)$ is the manifold formed by the set of initial conditions $\boldsymbol{x}_i(0)$ in the basin of attraction of the limit cycle, such that the trajectories leaving from $\boldsymbol{x}_i(0)$ are attracted to $\boldsymbol{x}_s(t)$, that is

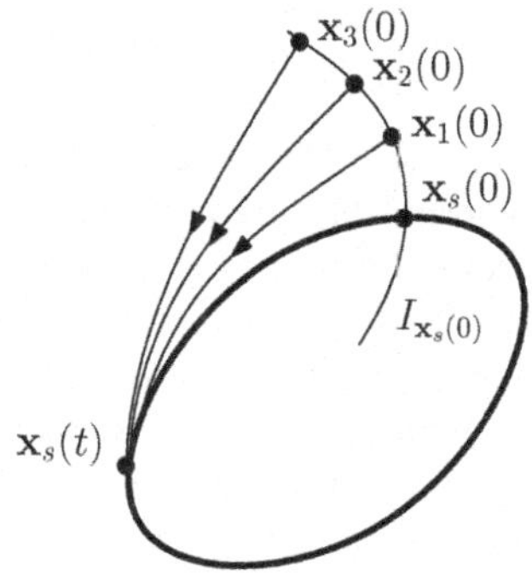

Fig. 11.3 Definition of an isochron

$$I_{x_s(0)} = \left\{ x_i(0) \in \mathbb{R}^n / x_s \mid \lim_{t \to +\infty} ||x_i(t) - x_s(t)|| = 0 \right\}. \tag{11.6}$$

Isochrons represent the stable manifold foliation of $x_s(t)$ (see Fig. 11.3). The phase of the points belonging to the basin of attraction of the stable limit cycle is defined assigning the same phase to points lying on the same isochron. In this way, the phase of the point $x_i(t)$ results to be $\phi(x_i(t)) = \omega t + \phi(x_i(0))$. Thus the isochrons are the level sets of the scalar field $\phi(x)$.

Corollary 1 *Isochrons are the invariant set for the Poincaré first return map* $\Phi : I_{x_s(t)} \mapsto I_{x_s(t+T)}$. *Different leaves are permuted by the flow, that is*

$$x_i(0) \in I_{x_s(0)} \Leftrightarrow x_i(t) \in I_{x_s(t)}$$

Proof It follows directly from the definition of isochron and from the definition of the phase function. □

Isochrons and the phase function proved valuable tools to understand the physics of weakly perturbed oscillators, and the mechanisms leading to synchronization of coupled oscillators, but unfortunately their application to make quantitative predictions about real world oscillators is hindered by the fact that they can rarely be found analytically. Thus, one has to resort to rather sophisticated algorithms for their calculation [94–100].

An alternative solution is to decompose the oscillator's dynamics into two components, one tangent and one transversal (but not necessarily orthogonal) to the limit cycle [91, 101–111]. The dynamics of the projection along the cycle represents the phase dynamics, whereas the transversal components are interpreted as the amplitude deviation. Obviously, both the phase and amplitude deviation variables depend upon the projection used, however it can be shown that, if the projection operators are chosen according to Floquet theory, in the neighborhood of the limit cycle the phase variable coincides with the phase defined using isochrons [104, 105].

Consider a network composed by N weakly coupled dynamical systems. The normalized state equations read

$$\frac{\mathrm{d}\boldsymbol{x}_i}{\mathrm{d}\tau} = \boldsymbol{a}_i(\boldsymbol{x}_i, \boldsymbol{\mu}_i) + \varepsilon \boldsymbol{c}_i(\boldsymbol{x}_1, \ldots, \boldsymbol{x}_N) \qquad \text{for } i = 1, \ldots, N \tag{11.7}$$

where $\boldsymbol{x}_i : \mathbb{R}^+ \mapsto \mathbb{R}^n$ and $\boldsymbol{a}_i : \mathbb{R}^n \mapsto \mathbb{R}^n$ are vector valued functions describing the state and the internal dynamics of the oscillators, respectively, $\boldsymbol{\mu}_i \in \mathbb{R}^m$ are parameter vectors, $\varepsilon \ll 1$ is a small parameter that defines the interaction strength, and $\boldsymbol{c}_i : \mathbb{R}^{n\times N} \mapsto \mathbb{R}^n$ are vector valued functions describing the interactions. We shall assume that all functions are sufficiently smooth.

Natural and even man made systems can never be perfectly identical. This can be accounted for assuming that the parameters $\boldsymbol{\mu}_i$ are subject to a small deviation from an average value $\boldsymbol{\mu}_i = \boldsymbol{\mu}_0 + \delta\boldsymbol{\mu}_i$. Expanding the vector field around $\boldsymbol{\mu}_0$

$$\boldsymbol{a}_i(\boldsymbol{x}_i, \boldsymbol{\mu}_i) = \boldsymbol{a}_i(\boldsymbol{x}_i, \boldsymbol{\mu}_0) + \frac{\partial \boldsymbol{a}_i(\boldsymbol{x}_i, \boldsymbol{\mu}_0)}{\partial \boldsymbol{\mu}_i}\delta\boldsymbol{\mu}_i + \cdots \tag{11.8}$$

where $\partial \boldsymbol{a}_i(\boldsymbol{x}_i, \boldsymbol{\mu}_0)/\partial\boldsymbol{\mu}$ is the matrix of partial derivatives of $\boldsymbol{a}_i$ with respect to $\boldsymbol{\mu}$. For the sake of simplicity, we shall assume that the dynamic equations of the uncoupled systems are identical, and that the differences are small enough to approximate (11.8) as

$$\boldsymbol{a}_i(\boldsymbol{x}_i, \boldsymbol{\mu}_i) \approx \boldsymbol{a}(\boldsymbol{x}_i) + \varepsilon \boldsymbol{p}_i(\boldsymbol{x}_i) \tag{11.9}$$

Thus, the governing equations (11.7) can be rewritten as

$$\frac{\mathrm{d}\boldsymbol{x}_i}{\mathrm{d}\tau} = \boldsymbol{a}(\boldsymbol{x}_i) + \varepsilon \boldsymbol{p}_i(\boldsymbol{x}_i) + \varepsilon \boldsymbol{c}_i(\boldsymbol{x}_1, \ldots, \boldsymbol{x}_N) \tag{11.10}$$

for $i = 1, \ldots, N$.

It is convenient to introduce an ideal dynamical system, considered as reference

$$\frac{\mathrm{d}\boldsymbol{x}}{\mathrm{d}\tau} = \boldsymbol{a}(\boldsymbol{x}) \tag{11.11}$$

It is important to stress that the reference system does not need to really exist, let alone belong to the considered set of oscillators. We restrict the attention to the case where the reference unit is an oscillator, i.e. we assume that (11.11) has an asymptotically stable T–periodic solution $\boldsymbol{x}_s(\tau)$ such that

$$\begin{cases} \dfrac{\mathrm{d}\boldsymbol{x}_s(\tau)}{\mathrm{d}\tau} = \boldsymbol{a}(\boldsymbol{x}_s(\tau)) \\ \boldsymbol{x}_s(\tau) = \boldsymbol{x}_s(\tau + T) \end{cases} \tag{11.12}$$

representing a limit cycle in the state space. The periodic solution is used to define the unit vector tangent to the limit cycle

$$\boldsymbol{u}_1(\tau) = \frac{\boldsymbol{a}(\boldsymbol{x}_s(\tau))}{|\boldsymbol{a}(\boldsymbol{x}_s(\tau))|} \tag{11.13}$$

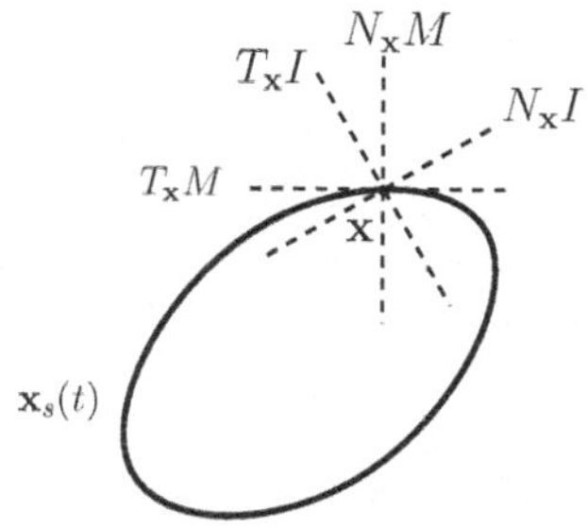

Fig. 11.4 State space decomposition for a second order oscillator, illustrating the tangent spaces $T_x M$, $T_x I$, and the normal spaces $N_x M$, $N_x I$

where $|\cdot|$ denotes the L_2 norm. Together with $\boldsymbol{u}_1(\tau)$, we consider other $n-1$ linear independent vectors $\boldsymbol{u}_2(\tau), \ldots, \boldsymbol{u}_n(\tau)$, such that the set $\{\boldsymbol{u}_1(\tau), \ldots, \boldsymbol{u}_n(\tau)\}$ is a basis for $\mathbb{R}^n$, for all τ. Given the matrix $\boldsymbol{U}(\tau) = [\boldsymbol{u}_1(\tau), \ldots, \boldsymbol{u}_n(\tau)]$, we define the reciprocal vectors $\boldsymbol{v}_1^T(\tau), \ldots, \boldsymbol{v}_n^T(\tau)$ to be the rows of the inverse matrix $\boldsymbol{V}(\tau) = \boldsymbol{U}^{-1}(\tau)$. Thus $\{\boldsymbol{v}_1(\tau), \ldots, \boldsymbol{v}_n(\tau)\}$ also spans $\mathbb{R}^n$ and the bi–orthogonality condition $\boldsymbol{v}_i^T \boldsymbol{u}_j = \boldsymbol{u}_i^T \boldsymbol{v}_j = \delta_{ij}$ for all τ, holds. We shall also use matrices $\boldsymbol{Y}(\tau) = [\boldsymbol{u}_2(\tau), \ldots, \boldsymbol{u}_n(\tau)]$, $\boldsymbol{Z}(\tau) = [\boldsymbol{v}_2(\tau), \ldots, \boldsymbol{v}_n(\tau)]$, and the magnitude of the vector field evaluated on the limit cycle, $r(\tau) = |\boldsymbol{a}(\boldsymbol{x}_s(\tau))|$.

The use of Floquet bases for the sets $\{\boldsymbol{u}_1(\tau), \ldots, \boldsymbol{u}_n(\tau)\}$ and $\{\boldsymbol{v}_1(\tau), \ldots, \boldsymbol{v}_n(\tau)\}$ turns out to be very convenient, because they realize a complete decomposition of the state space. At each point $\boldsymbol{x} \in \boldsymbol{x}_s$, the limit cycle's stable manifold can be decomposed into two complementary linear spaces, the space $T_x M$ tangent to the limit cycle at $\boldsymbol{x}$, and the space $T_x I$ tangent to the isochron at $\boldsymbol{x}$. Extending the definition to all points belonging on the limit cycle leads to the tangent bundles TM and TI, that are spanned by the vectors $\boldsymbol{u}_1(\tau)$ and $\{\boldsymbol{u}_2(\tau), \ldots, \boldsymbol{u}_n(\tau)\}$, respectively. The covector $\boldsymbol{v}_1(\tau)$ is normal to all vectors $\boldsymbol{u}_k(\tau)$, for $k = 2, \ldots, n$ and in turn, each covector $\boldsymbol{v}_k(\tau)$, $k = 2, \ldots, n$ is normal to $\boldsymbol{u}_1(\tau)$. Thus $\boldsymbol{v}_1(\tau)$ spans the one-dimensional cotangent bundle NI orthogonal to TI, while the covectors $\{\boldsymbol{v}_2(\tau), \ldots, \boldsymbol{v}_n(\tau)\}$ span the $(n-1)$-dimensional cotangent bundle NM orthogonal to TM [112, 113]. The picture for a second order oscillator is illustrated in Fig. 11.4.

The following theorem establishes the equations governing the time evolution for the phase and amplitude deviation variables.

Theorem 1 (Phase-amplitude deviation equations) *Consider system (11.10), admitting of a T-periodic limit cycle $\boldsymbol{x}_s(\tau)$ for $\varepsilon = 0$. Let $\{\boldsymbol{u}_1(\tau), \ldots, \boldsymbol{u}_n(\tau)\}$ and $\{\boldsymbol{v}_1(\tau), \ldots, \boldsymbol{v}_n(\tau)\}$ be two reciprocal bases such that $\boldsymbol{u}_1(\tau)$ satisfies (11.13) and such that the bi–orthogonality condition $\boldsymbol{v}_i^T \boldsymbol{u}_j = \boldsymbol{u}_i^T \boldsymbol{v}_j = \delta_{ij}$ holds. Consider the coordinate transformation (dependence on τ in θ_i and $\boldsymbol{R}_i$ is omitted for simplicity of notation)*

$$\boldsymbol{x}_i(\tau) = \boldsymbol{h}(\theta_i, \boldsymbol{R}_i) = \boldsymbol{x}_s(\theta_i) + \boldsymbol{Y}(\theta_i)\boldsymbol{R}_i \tag{11.14}$$

Then a neighborhood of the limit cycle $\boldsymbol{x}_s(\tau)$ exists where the phase $\theta_i(\tau)$ satisfies

$$\frac{d\theta_i}{d\tau} = 1 + a_\theta(\theta_i, \boldsymbol{R}_i) + \varepsilon\, \delta a_{\theta_i}(\theta_i, \boldsymbol{R}_i) + \varepsilon C_{\theta_i}(\theta_1, \boldsymbol{R}_1, \ldots, \theta_N, \boldsymbol{R}_N) \tag{11.15}$$

with

$$K(\theta_i, \boldsymbol{R}_i) = \left(|\boldsymbol{a}(\boldsymbol{x}_s(\theta_i))| + \boldsymbol{v}_1^T(\theta_i) \frac{\partial \boldsymbol{Y}(\theta_i)}{\partial \theta_i} \boldsymbol{R}_i \right)^{-1} \tag{11.16a}$$

$$a_\theta(\theta_i, \boldsymbol{R}_i) = K(\theta_i, \boldsymbol{R}_i)\, \boldsymbol{v}_1^T(\theta_i) \left(\boldsymbol{a}(\boldsymbol{x}_s(\theta_i) + \boldsymbol{Y}(\theta_i)\boldsymbol{R}_i) - \boldsymbol{a}(\boldsymbol{x}_s(\theta_i)) - \frac{\partial \boldsymbol{Y}(\theta_i)}{\partial \theta_i} \boldsymbol{R}_i \right) \tag{11.16b}$$

$$\delta a_{\theta_i}(\theta_i, \boldsymbol{R}_i) = K(\theta_i, \boldsymbol{R}_i)\, \boldsymbol{v}_1^T(\theta_i)\, \boldsymbol{p}_i\big(\boldsymbol{x}_s(\theta_i) + \boldsymbol{Y}(\theta_i)\boldsymbol{R}_i\big) \tag{11.16c}$$

$$\begin{aligned} &C_{\theta_i}(\theta_1, \boldsymbol{R}_1, \ldots, \theta_N, \boldsymbol{R}_N) \\ &\quad = K(\theta_i, \boldsymbol{R}_i)\, \boldsymbol{v}_1^T(\theta_i)\, \boldsymbol{c}_i\big(\boldsymbol{x}_s(\theta_1) + \boldsymbol{Y}(\theta_1)\boldsymbol{R}_1, \ldots, \boldsymbol{x}_s(\theta_N) + \boldsymbol{Y}(\theta_N)\boldsymbol{R}_N\big) \end{aligned} \tag{11.16d}$$

The amplitude deviations satisfy

$$\frac{d\boldsymbol{R}_i}{d\tau} = \boldsymbol{L}(\theta_i)\boldsymbol{R}_i + \boldsymbol{a}_{\boldsymbol{R}}(\theta_i, \boldsymbol{R}_i) + \varepsilon\, \delta \boldsymbol{a}_{\boldsymbol{R}_i}(\theta_i, \boldsymbol{R}_i) + \varepsilon\, \boldsymbol{C}_{\boldsymbol{R}_i}(\theta_1, \boldsymbol{R}_1, \ldots, \theta_N, \boldsymbol{R}_N) \tag{11.17}$$

where

$$\boldsymbol{L}(\theta_i) = -\boldsymbol{Z}^T(\theta_i) \frac{\partial \boldsymbol{Y}(\theta_i)}{\partial \theta_i} \tag{11.18a}$$

$$\boldsymbol{a}_{\boldsymbol{R}}(\theta_i, \boldsymbol{R}_i) = -\boldsymbol{Z}^T(\theta_i) \left(\frac{\partial \boldsymbol{Y}(\theta_i)}{\partial \theta_i} \boldsymbol{R}_i\, a_\theta(\theta_i, \boldsymbol{R}_i) - \boldsymbol{a}\big(\boldsymbol{x}_s(\theta_i) + \boldsymbol{Y}(\theta_i)\boldsymbol{R}_i\big) \right) \tag{11.18b}$$

$$\delta \boldsymbol{a}_{\boldsymbol{R}_i}(\theta_i, \boldsymbol{R}_i) = -\boldsymbol{Z}^T(\theta_i) \left(\frac{\partial \boldsymbol{Y}(\theta_i)}{\partial \theta_i} \boldsymbol{R}_i\, \delta a_\theta(\theta_i, \boldsymbol{R}_i) - \boldsymbol{p}_i\big(\boldsymbol{x}_s(\theta_i) + \boldsymbol{Y}(\theta_i)\boldsymbol{R}_i\big) \right) \tag{11.18c}$$

$$\begin{aligned} \boldsymbol{C}_{\boldsymbol{R}_i}(\theta_1, \boldsymbol{R}_1, \ldots, \theta_N, \boldsymbol{R}_N) &= -\boldsymbol{Z}^T(\theta_i) \frac{\partial \boldsymbol{Y}(\theta_i)}{\partial \theta_i} C_{\theta_i}(\theta_1, \boldsymbol{R}_1, \ldots, \theta_N, \boldsymbol{R}_N)\boldsymbol{R}_i \\ &\quad + \boldsymbol{Z}^T(\theta_i)\, \boldsymbol{c}_i\big(\boldsymbol{x}_s(\theta_1) + \boldsymbol{Y}(\theta_1)\boldsymbol{R}_1, \ldots, \boldsymbol{x}_s(\theta_N) + \boldsymbol{Y}(\theta_N)\boldsymbol{R}_N\big) \end{aligned} \tag{11.18d}$$

Proof First, we show that the coordinate transformation (11.14) is invertible in the neighborhood of the limit cycle, and thus the problem is well posed. The Jacobian matrix of the coordinate transformation is

$$D\boldsymbol{h}(\theta_i, \boldsymbol{R}_i) = \left[\frac{\partial \boldsymbol{h}}{\partial \theta_i} \quad \frac{\partial \boldsymbol{h}}{\partial \boldsymbol{R}_i} \right] = \left[\frac{\partial \boldsymbol{x}_s(\theta_i)}{\partial \theta_i} + \frac{\partial \boldsymbol{Y}(\theta_i)}{\partial \theta_i} \boldsymbol{R}_i \quad \boldsymbol{Y}(\theta_i) \right] \tag{11.19}$$

On the limit cycle $\boldsymbol{R}_i = 0$ and then

$$D\boldsymbol{h}(\theta_i, \boldsymbol{R}_i)\big|_{\boldsymbol{R}_i=0} = \left[\frac{\partial \boldsymbol{x}_s(\theta_i)}{\partial \theta_i} \quad \boldsymbol{Y}(\theta_i) \right] = \big[|\boldsymbol{a}\big(\boldsymbol{x}_s(\theta_i)\big)|\, \boldsymbol{u}_1(\theta_i), \boldsymbol{u}_2(\theta_i), \ldots, \boldsymbol{u}_n(\theta_i) \big]$$

Since $\{\boldsymbol{u}_1(\tau), \ldots, \boldsymbol{u}_n(\tau)\}$ is a basis for $\mathbb{R}^n$, it follows that the determinant of the Jacobian matrix is not zero. Then by the inverse function theorem there exists a neighborhood of $\boldsymbol{R}_i = 0$ where $\boldsymbol{h}$ is invertible. Moreover, if $\boldsymbol{h}$ is of class C^k then its inverse is also of class C^k.

Next we derive (11.15) and (11.17). Taking the derivative of (11.14) we have

$$\begin{aligned}\frac{\mathrm{d}\boldsymbol{x}_i}{\mathrm{d}\tau} &= \frac{\partial \boldsymbol{h}(\theta_i, \boldsymbol{R}_i)}{\partial \theta_i}\frac{\mathrm{d}\theta_i}{\mathrm{d}\tau} + \frac{\partial \boldsymbol{h}(\theta_i, \boldsymbol{R}_i)}{\partial \boldsymbol{R}_i}\frac{\mathrm{d}\boldsymbol{R}_i}{\mathrm{d}\tau} \\ &= \left(\frac{\partial \boldsymbol{x}_\mathrm{s}(\theta_i)}{\partial \theta_i} + \frac{\partial \boldsymbol{Y}(\theta_i)}{\partial \theta_i}\boldsymbol{R}_i\right)\frac{\mathrm{d}\theta_i}{\mathrm{d}\tau} + \boldsymbol{Y}(\theta_i)\frac{\mathrm{d}\boldsymbol{R}_i}{\mathrm{d}\tau} = \\ &= \boldsymbol{a}\Big(\boldsymbol{x}_\mathrm{s}(\theta_i) + \boldsymbol{Y}(\theta_i)\boldsymbol{R}_i\Big) + \varepsilon \boldsymbol{p}_i\Big(\boldsymbol{x}_\mathrm{s}(\theta_i) + \boldsymbol{Y}(\theta_i)\boldsymbol{R}_i\Big) \\ &\quad + \varepsilon \boldsymbol{c}_i\Big(\boldsymbol{x}_\mathrm{s}(\theta_1) + \boldsymbol{Y}(\theta_1)\boldsymbol{R}_1, \ldots, \boldsymbol{x}_\mathrm{s}(\theta_N) + \boldsymbol{Y}(\theta_N)\boldsymbol{R}_N\Big) \end{aligned} \tag{11.20}$$

Multiplying to the left by $\boldsymbol{v}_1(\theta_i)^T$, using the bi-orthogonality condition and rearranging terms we obtain (11.15). Conversely, multiplying to the left by $\boldsymbol{Z}^T(\theta_i)$ we find (11.17). □

Functions $\theta_i : \mathbb{R}^+ \mapsto \mathbb{R}^+$ can be interpreted as an elapsed time from the initial condition, representing a new parametrization of the trajectories. Because they are measured on the limit cycle, they map the interval $[0, T) \rightarrow [0, 2\pi)$ and can be viewed as *phase functions*. Functions $\boldsymbol{R}_i : \mathbb{R}^+ \mapsto \mathbb{R}^{n-1}$ represent orbital deviations from the limit cycle, and they will be referred to as the *amplitude deviation functions* or *amplitudes*, for simplicity.

Figure 11.5 shows the idea behind this decomposition for a second order system. At any time instant the solution of the perturbed ($\varepsilon \neq 0$, blue line) system $\boldsymbol{x}_i(t)$ can be decomposed into two contributions. The first component is the unperturbed limit cycle $\boldsymbol{x}_\mathrm{s}(\theta_i(t))$ evaluated at the unknown time instant $\theta_i(t)$, plus a distance $\boldsymbol{R}_i(t)$ measured along the directions spanned by the vectors $\boldsymbol{u}_2, \ldots, \boldsymbol{u}_n$.

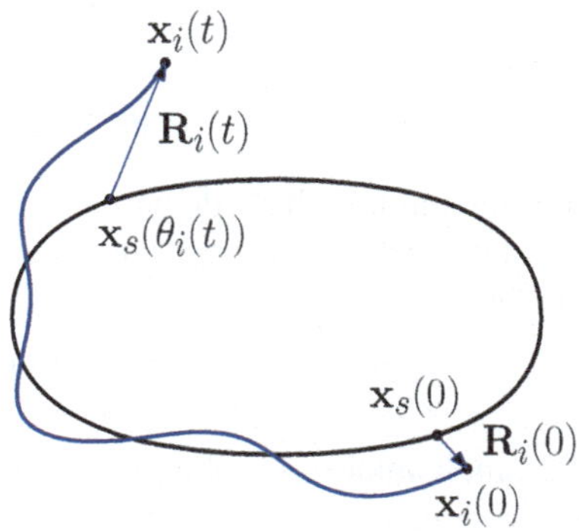

Fig. 11.5 State space decomposition of the solution $\boldsymbol{x}_i(t)$

For most practical applications, the phase is the most important variable. In fact, perturbations along the cycle are neither absorbed nor amplified, since this variable corresponds to the unitary Floquet multiplier and thus to the sole neutrally stable direction. They persist and become eventually unbounded asymptotically with time. Conversely, perturbations transverse to the cycle are absorbed, being associated to the remaining Floquet multipliers (those with magnitude less than one for a stable limit cycle), as proved in the following corollary.

Corollary 2 *Consider the reference oscillator (11.11) with the T-periodic limit cycle $\boldsymbol{x}_s(\tau)$. Let $1, \mu_2, \ldots, \mu_n$ be the characteristic multipliers of the variational equation*

$$\frac{d\boldsymbol{y}(\tau)}{d\tau} = \frac{\partial \boldsymbol{a}(\boldsymbol{x}_s(\tau))}{\partial \boldsymbol{x}} \boldsymbol{y}(\tau) \tag{11.21}$$

Then $\boldsymbol{R}_i = 0$ is an equilibrium point for the amplitude equation of the unperturbed system,

$$\frac{d\boldsymbol{R}_i}{dt} = \boldsymbol{L}(\theta_i)\boldsymbol{R}_i + \boldsymbol{a}_{\boldsymbol{R}}(\theta_i, \boldsymbol{R}_i) \tag{11.22}$$

with characteristic multipliers $\mu_2, \ldots, \mu_n$.

Proof See [104], theorem 2. □

The far reaching consequence of Corollary 2 is that, for weakly coupled oscillators that possess strongly attractive limit cycles, the phase dynamics can be effectively decoupled from the amplitude dynamics. In fact, for strongly stable limit cycles, it is reasonable to assume that the amplitude deviations remain small, so that we can assume $\boldsymbol{R}_i \approx 0$ and simplify (11.15) into

$$\frac{\mathrm{d}\theta_i}{\mathrm{d}\tau} = 1 + \varepsilon \delta a_{\theta_i}(\theta_i) + \varepsilon C_{\theta_i}(\theta_1 \ldots, \theta_N) \tag{11.23}$$

where

$$\delta a_{\theta_i}(\theta_i) = \frac{1}{|\boldsymbol{a}(\boldsymbol{x}_\mathrm{s}(\theta_i))|} \boldsymbol{v}_i^T(\theta_i)\, \boldsymbol{p}_i\Big(\boldsymbol{x}_\mathrm{s}(\theta_i)\Big) \tag{11.24}$$

$$C_{\theta_i}(\theta_1, \ldots, \theta_N) = \frac{1}{|\boldsymbol{a}(\boldsymbol{x}_\mathrm{s}(\theta_i))|} \boldsymbol{v}_i^T(\theta_i)\, \boldsymbol{c}_i\Big(\boldsymbol{x}_\mathrm{s}(\theta_1), \ldots, \boldsymbol{x}_\mathrm{s}(\theta_N)\Big) \tag{11.25}$$

Introducing the phase deviation $\psi_i = \theta_i - \tau$, representing the variation of the i-th oscillator phase with respect to the phase of the reference, and using (11.23), we obtain

$$\frac{\mathrm{d}\psi_i}{\mathrm{d}\tau} = \varepsilon\, \delta a_{\theta_i}(\psi_i + \tau) + \varepsilon\, C_{\theta_i}(\psi_1 + \tau, \ldots, \psi_N + \tau) \tag{11.26}$$

For small values of ε, the phase deviation is therefore a slow variable. We can thus average the right-hand-side integrating with respect to τ from 0 to $T = 2\pi$ without introducing a significant error, obtaining the *phase deviation equation*

$$\frac{d\psi_i}{d\tau} = \varepsilon\,\overline{\delta a}_{\theta_i}(\psi_i) + \varepsilon\,\overline{C}_{\theta_i}(\psi_1, \ldots, \psi_N) \quad \text{for } i = 1, \ldots, N \tag{11.27}$$

where

$$\overline{\delta a}_{\theta_i}(\psi_i) = \frac{1}{2\pi}\int_0^{2\pi} \delta a_{\theta_i}(\psi_i + \tau)d\tau \tag{11.28}$$

$$\overline{C}_{\theta}(\psi_1, \ldots, \psi_N) = \frac{1}{2\pi}\int_0^{2\pi} C_{\theta_i}(\psi_1 + \tau, \ldots, \psi_N + \tau)\,d\tau \tag{11.29}$$

Equilibrium points of the averaged phase deviation Eq. (11.27) corresponds to phase locked oscillations in the network of coupled oscillators.

Thus, (11.27) can be used to design networks that exploit phase locked oscillations to perform fundamental logical operations.

11.3.1 Example

Before we consider the use (11.27) to design logic gates, we give an example of phase equation reduction at the single oscillator level. This will also serve as the starting point for further analysis.

Consider the nonlinear oscillator described by (11.5), that we rewrite in the form

$$\frac{dx}{d\tau} = y \tag{11.30a}$$

$$\frac{dy}{d\tau} = -x + \alpha(1 - y^2)y \tag{11.30b}$$

where we have assumed $G_1 = G_3 = \alpha$.

For second order systems, Floquet basis and co-basis can be computed using the formulas given in [84], while for higher order systems, numerical methods are needed [114–116]. Since the goal of this work is to illustrate fundamental principles avoiding technicalities as much as possible, we shall consider the weakly nonlinear version of (11.30) (i.e., we assume $\alpha \ll 1$). This permits to introduce some approximations allowing for the analytical determination of Floquet vectors and co-vectors. Moreover, these vectors take a particularly simple form that greatly simplifies the following discussion.

Introducing polar coordinates $x = A\cos\theta$, $y = A\sin\theta$, and averaging on the interval $[0, 2\pi[$, (11.30) becomes

$$\frac{d\theta}{d\tau} = 1 \tag{11.31a}$$

$$\frac{dA}{d\tau} = \frac{\alpha}{2} A \left(1 - \frac{3}{4} A^2\right) \tag{11.31b}$$

Equation (11.31) admits the asymptotic solution

$$\boldsymbol{x}_s(\tau) = \begin{pmatrix} \theta(\tau) \\ A(\tau) \end{pmatrix} = \begin{pmatrix} \tau \\ 2/\sqrt{3} \end{pmatrix} \tag{11.32}$$

Defining $\theta(\tau) = \tau \mod (2\pi)$, the solution is periodic and defines a limit cycle. The Jacobian matrix evaluated in the solution reads

$$J|_{\boldsymbol{x}_s} = \begin{pmatrix} 0 & 0 \\ 0 & -\alpha \end{pmatrix} \tag{11.33}$$

The eigenvalues of the Jacobian matrix, $\lambda_1 = 0$ and $\lambda_2 = -\alpha$, are the cycle Floquet exponents. As expected, one exponent is null, the so called *structural exponent*, while the second one is negative, implying that the limit cycle is asymptotically stable. The eigenvectors of the Jacobian matrix $\boldsymbol{u}_1 = [1, 0]^T$ and $\boldsymbol{u}_2 = [0, 1]^T$, are the associated Floquet vectors. They form an orthogonal unitary basis, thus the co-vectors are $\boldsymbol{v}_1 = \boldsymbol{u}_1$ and $\boldsymbol{v}_2 = \boldsymbol{u}_2$. The Floquet vectors $\boldsymbol{u}_1$, $\boldsymbol{u}_2$ and co-vectors $\boldsymbol{v}_1$, $\boldsymbol{v}_2$ can be used to derive the amplitude-phase model and the phase deviation equation for a coupled network.

11.4 Oscillator Networks for Boolean Logic

Boolean logic is based on binary variable states, let them be the ON and OFF states in electronic systems, 0 and 1 in computer science or $|\uparrow\rangle$ and $|\downarrow\rangle$ in quantum mechanics-based computations. In a coupled oscillators network the binary states correspond to in-phase and anti-phase locked oscillations, meaning that an oscillator must have the same phase (*in-phase locking*), or have a phase difference equal to π (*anti-phase locking*) with respect to the reference oscillator (Fig. 11.6).

11.4.1 Registers

Information is stored into *registers,* represented by oscillators connected to the central reference unit. As stated, the connection is one-directional, that is, the reference influences the registers, but not vice versa. By a proper engineering of the coupling, a register can be in-phase, or anti-phase, locked with the reference signal, corresponding to a stored datum equal to 0 or 1, respectively.

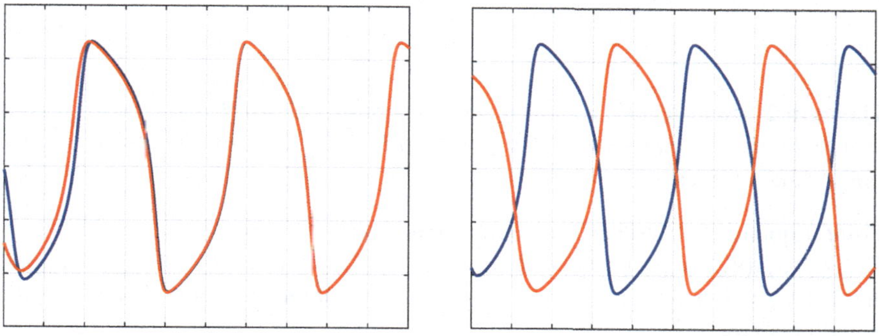

Fig. 11.6 Oscillator waveforms exhibiting phase locking after a short transient. On the left: In-phase locked waveforms. On the right: Anti-phase locked waveforms

Fig. 11.7 Realization of the reference-register. Controlled sources represent the reference

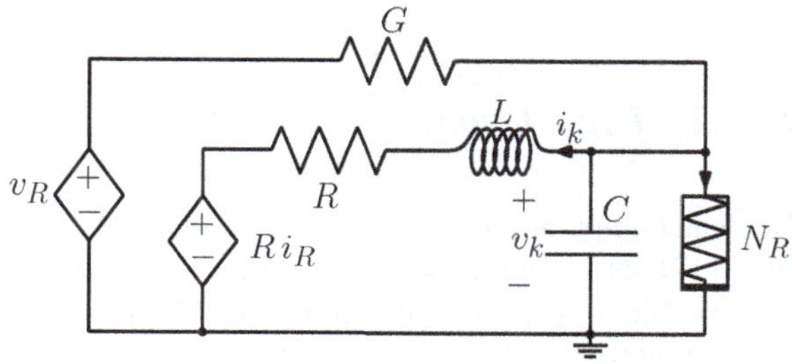

Figure 11.7 shows a possible realization of a reference-register structure. The reference oscillator is represented by the controlled sources. Applying Kirchhoff laws the following state equations are obtained

$$\frac{di_k}{dt} = \frac{1}{L}v_k - \frac{R}{L}(i_k + i_R) \tag{11.34a}$$

$$\frac{dv_k}{dt} = -\frac{1}{C}i_k - \frac{1}{C}g(v_k) - \frac{G}{C}(v_k - v_R) \tag{11.34b}$$

where v_R and i_R are the state variables of the reference, while v_k and i_k refer to the k-th register. We have also assumed that all oscillators are characterized by the same nonlinear conductance $g(v_k)$. After proper normalization

$$\frac{dx_k}{d\tau} = y_k - 2\rho(x_k + x_R) \tag{11.35a}$$

$$\frac{dy_k}{d\tau} = -x_k - g(y_k) - 2\gamma(y_k - y_R) \tag{11.35b}$$

where $2\rho = R\sqrt{C/L}$ and $2\gamma = G\sqrt{L/C}$. Applying the procedure described in Sect. 11.3 we obtain the phase deviation equation

$$\frac{d\psi_k}{d\tau} = (\rho - \gamma)\sin\psi_k \tag{11.36}$$

Theorem 2 (Register) *Consider the phase equation (11.36). For $\gamma > \rho$, $\overline{\psi}_k = 0$ is an asymptotically stable equilibrium point, while $\overline{\psi}_k = \pi$ is unstable. For $\rho > \gamma$ the situation is reversed.*

Proof Equation (11.36) has two equilibrium points, $\overline{\psi}_k = 0$ and $\overline{\psi}_k = \pi$.
For $\rho > \gamma$, $d\psi_k/d\tau > 0$ for $0 < \psi < \pi$ and $d\psi_k/d\tau < 0$ for $\pi < \psi < 2\pi$, implying that $\overline{\psi}_k = 0$ is unstable, while $\overline{\psi}_k = \pi$ is asymptotically stable. For $\rho < \gamma$ the situation is reversed. □

Therefore, controlling the value of parameters ρ and γ it is possible to switch the state of the register between the two digital values.

11.4.2 Logic Gates

Boolean computation requires the availability of a *logically* or *functionally complete* [67, 117] set of elementary logical operations that can be composed to implement any combinational logic function. When logic states are encoded into phases, it is convenient to use the set composed by the NOT and the MAJORITY gates [67].

11.4.2.1 NOT Gate

A NOT gate is a single input, single output gate, that implements logical negation. An example of coupled oscillators realizing an invertible NOT gate is shown in Fig. 11.8.

Two oscillators, denoted as j and k, are coupled through resistor R. One oscillator represents the input, while the other is the output. The gate is denoted as *reversible* because the input and output oscillators can be interchanged without impairing the operation. For instance, in Fig. 11.8 oscillator j is the input while oscillator k is

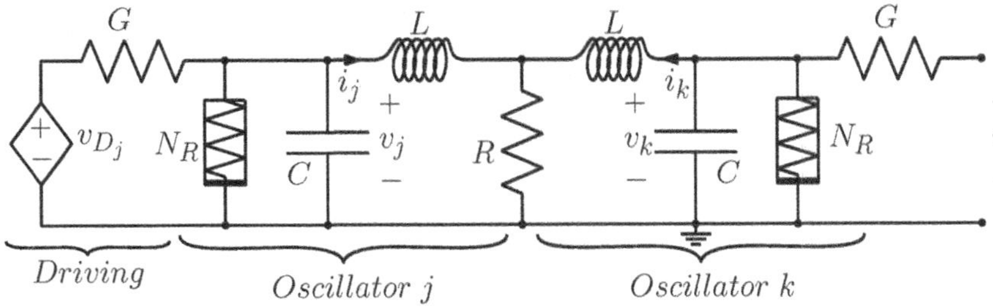

Fig. 11.8 Reversible NOT gate. The controlled voltage sources represent the driving signal used to select the input state

the output. The state of the input is controlled through the coupling with a driving oscillator unit, here represented by the controlled voltage source v_{D_j}, while the output is voltage $v_{D_k} = v_k$.

Applying KVL and KCL, and proper normalization, the state equations become

$$\frac{dx_j}{d\tau} = y_j - 2\rho(x_j + x_k) \tag{11.37a}$$

$$\frac{dy_j}{d\tau} = -x_j - g(y_j) - 2\gamma(y_j - y_{D_j}) \tag{11.37b}$$

$$\frac{dx_k}{d\tau} = y_k - 2\rho(x_k + x_j) \tag{11.37c}$$

$$\frac{dy_k}{d\tau} = -x_k - g(y_k) \tag{11.37d}$$

The corresponding phase deviation equations read

$$\frac{d\psi_j}{d\tau} = \rho \sin(\psi_j - \psi_k) - \gamma \sin(\psi_j - \psi_{D_j}) \tag{11.38a}$$

$$\frac{d\psi_k}{d\tau} = \rho \sin(\psi_k - \psi_j) \tag{11.38b}$$

The following theorem proves that the phase deviation equations (11.38) corresponds to a NOT gate.

Theorem 3 (NOT gate) *Consider the phase deviation Eq. (11.38), with $\rho > 0$ and $\gamma > 0$. Let $\psi_{D_j} = 0$ (respectively $\psi_{D_j} = \pi$), then $(\overline{\psi}_j, \overline{\psi}_k) = (0, \pi)$ (respectively $(\overline{\psi}_j, \overline{\psi}_k) = (\pi, 0)$) is an asymptotically stable equilibrium point.*

Proof It is trivial to observe that $(\overline{\psi}_j, \overline{\psi}_k) = (0, \pi)$ and $(\overline{\psi}_j, \overline{\psi}_k) = (\pi, 0)$ are equilibrium points of (11.38). The proof is based on the idea to show the existence of a strict Liapunov function in a neighborhood of the equilibrium point. The existence of such a Liapunov function implies that the equilibrium point is asymptotically stable.

First, we consider the case $\psi_{D_j} = 0$, with equilibrium point $(\overline{\psi}_j, \overline{\psi}_k) = (0, \pi)$. Expanding one of the sine functions, we rewrite (11.38) in the form

$$\frac{d\psi_j}{d\tau} = \rho \sin(\psi_j - \psi_k) - \gamma \cos\psi_{D_j} \sin\psi_j \tag{11.39a}$$

$$\frac{d\psi_k}{d\tau} = \rho \sin(\psi_k - \psi_j) \tag{11.39b}$$

Consider the function $V : [0, 2\pi[\times[0, 2\pi[\mapsto \mathbb{R}$

$$V(\psi_j, \psi_k) = -\gamma \cos\psi_{D_j} \left(\cos\psi_j - 1\right) + \rho\left(\cos(\psi_j - \psi_k) + 1\right) \tag{11.40}$$

By definition, $V(\overline{\psi}_m, \overline{\psi}_n) = 0$, and $V(\psi_j, \psi_k) > 0$ for all ψ_j, ψ_k in a punctured neighborhood of $(\overline{\psi}_j, \overline{\psi}_k) = (0, \pi)$.

Finally, using definition (11.39), we have

$$\frac{\mathrm{d}V}{\mathrm{d}\tau} = \frac{\partial V}{\partial \psi_j}\frac{\mathrm{d}\psi_j}{\mathrm{d}\tau} + \frac{\partial V}{\partial \psi_k}\frac{\mathrm{d}\psi_k}{\mathrm{d}\tau} = -\left(\frac{\mathrm{d}\psi_j}{\mathrm{d}\tau}\right)^2 - \left(\frac{\mathrm{d}\psi_k}{\mathrm{d}\tau}\right)^2 < 0 \tag{11.41}$$

which proves that $V(\psi_j, \psi_k)$ is a strict Liapunov function in a neighborhood of $(0, \pi)$, as required.

The case $\psi_{D_j} = \pi$, with equilibrium point $(\overline{\psi}_j, \overline{\psi}_k) = (\pi, 0)$, is proved similarly, showing that the function

$$V(\psi_j, \psi_k) = -\gamma \cos \psi_{D_j} \left(\cos \psi_j + 1\right) + \rho \left(\cos(\psi_j - \psi_k) + 1\right) \tag{11.42}$$

is a strict Liapunov function in the neighborhood of $(\pi, 0)$. □

11.4.2.2 AND Gate and OR Gate

A MAJORITY gate is a three inputs, one output logic gate that implements both the AND and the OR logic operations according to the truth table shown in table 11.1.

An example of reversible MAJORITY gate implementation based on coupled oscillators is shown in Fig. 11.9. The first input is represented by the controlled voltage source v_D, which is applied to all oscillators and is used to set the function implemented by the gate. If the driving oscillator is set to zero (v_D is locked in-phase with the reference, in this case realized by the same driving circuit), then the MAJORITY gate realizes an AND gate. Conversely, if the driving register is set to one (v_D is anti-phase locked with the driving oscillator, this is realized applying a NOT to the reference), then the MAJORITY gate realizes an OR gate. This MAJORITY gate is called reversible because any pair of oscillators, i.e. (i, j), (i, k), or (j, k), can be used as input, with the remaining oscillator representing the output. In this

Table 11.1 Truth table for the MAJORITY gate

	Driving	Input 1	Input 2	Output
AND	0	0	0	0
	0	1	0	0
	0	0	1	0
	0	1	1	1
OR	1	0	0	0
	1	1	0	1
	1	0	1	1
	1	1	1	1

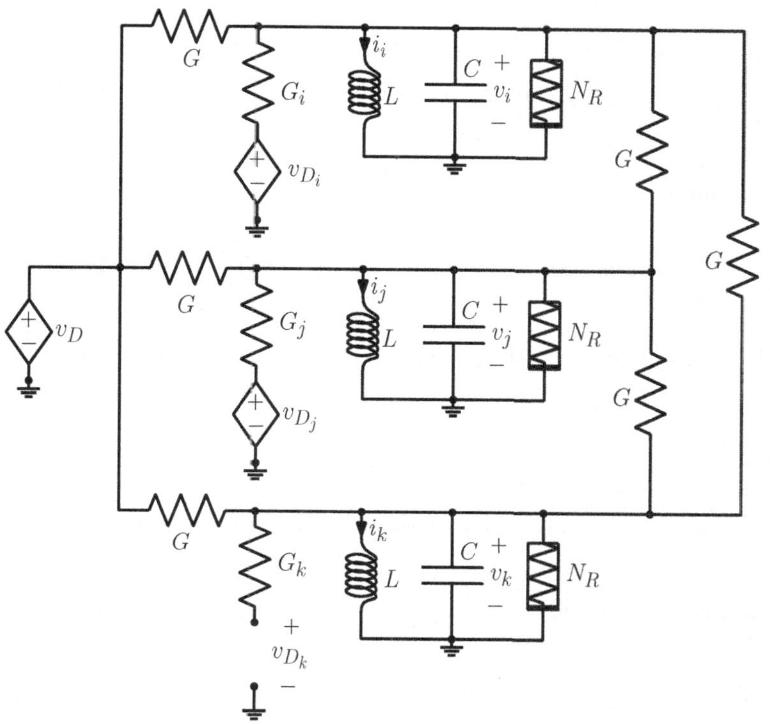

Fig. 11.9 Reversible MAJORITY gate. The controlled voltage sources represent the driving signals used to set the input state

example, the pair (i, j) defines the inputs, whilst k is the output. The input states are established applying driving signals v_{D_i} and v_{D_j} that can be set to zero (in-phase locking with the reference) or one (anti-phase locking). Voltage $v_{D_k} = v_k$ represents the output.

Applying KVL and KCL, and with proper normalization, the state equations become

$$\frac{\mathrm{d}x_i}{\mathrm{d}\tau} = y_i \tag{11.43a}$$

$$\frac{\mathrm{d}y_i}{\mathrm{d}\tau} = -x_i - g(x_i) - 2\gamma_i(y_i - y_{D_i}) - 2\gamma \sum_{m=j,k} (y_i - y_m) - 2\gamma(y_i - y_D) \tag{11.43b}$$

$$\frac{\mathrm{d}x_j}{\mathrm{d}\tau} = y_j \tag{11.43c}$$

$$\frac{\mathrm{d}y_j}{\mathrm{d}\tau} = -x_j - g(x_j) - 2\gamma_j(y_j - y_{D_j}) - 2\gamma \sum_{m=i,k} (y_j - y_m) - 2\gamma(y_j - y_D) \tag{11.43d}$$

$$\frac{dx_k}{d\tau} = y_k \tag{11.43e}$$

$$\frac{dy_k}{d\tau} = -x_k - g(x_k) - 2\gamma \sum_{m=i,j} (y_k - y_m) - 2\gamma (y_k - y_D) \tag{11.43f}$$

The resulting phase deviation equations read

$$\frac{d\psi_i}{d\tau} = -\gamma_i \sin(\psi_i - \psi_{D_i}) - \gamma \sin(\psi_i - \psi_D) - \gamma \sum_{m=j,k} \sin(\psi_i - \psi_m) \tag{11.44a}$$

$$\frac{d\psi_j}{d\tau} = -\gamma_j \sin(\psi_j - \psi_{D_j}) - \gamma \sin(\psi_j - \psi_D) - \gamma \sum_{m=i,k} \sin(\psi_j - \psi_m) \tag{11.44b}$$

$$\frac{d\psi_k}{d\tau} = -\gamma \sin(\psi_k - \psi_D) - \gamma \sum_{m=i,j} \sin(\psi_k - \psi_m) \tag{11.44c}$$

We are now in the position to prove that (11.44) realize both an AND and an OR gate.

Theorem 4 (AND gate) *Consider the phase deviation equation (11.44), with* $\psi_D = 0$ *and* $\gamma_i = \gamma_j > \gamma > 0$. *Then:*

(a) *If* $\psi_{D_i} = \psi_{D_j} = 0$, *then* $(\overline{\psi}_i, \overline{\psi}_j, \overline{\psi}_k) = (0, 0, 0)$ *is an asymptotically stable equilibrium point.*
(b) *If* $\psi_{D_i} = \pi$, $\psi_{D_j} = 0$ *(respectively* $\psi_{D_i} = 0$, $\psi_{D_j} = \pi$*), then* $(\overline{\psi}_i, \overline{\psi}_j, \overline{\psi}_k) = (\pi, 0, 0)$ *(respectively* $(\overline{\psi}_i, \overline{\psi}_j, \overline{\psi}_k) = (0, \pi, 0)$*) is an asymptotically stable equilibrium point.*
(c) *If* $\psi_{D_i} = \psi_{D_j} = \pi$ *then* $(\overline{\psi}_i, \overline{\psi}_j, \overline{\psi}_k) = (\pi, \pi, \pi)$ *is an asymptotically stable equilibrium point.*

Proof Rewrite the phase deviation equations as

$$\frac{d\psi_i}{d\tau} = -(\gamma_i \cos\psi_{D_i} + \gamma \cos\psi_D) \sin\psi_i - \gamma \left(\sin(\psi_i - \psi_j) + \sin(\psi_i - \psi_k)\right) \tag{11.45a}$$

$$\frac{d\psi_j}{d\tau} = -(\gamma_j \cos\psi_{D_j} + \gamma \cos\psi_D) \sin\psi_j - \gamma \left(\sin(\psi_j - \psi_i) + \sin(\psi_j - \psi_k)\right) \tag{11.45b}$$

$$\frac{d\psi_k}{d\tau} = -\gamma \cos\psi_D \sin\psi_k - \gamma \left(\sin(\psi_k - \psi_j) + \sin(\psi_k - \psi_j)\right) \tag{11.45c}$$

Consider the function $V : [0, 2\pi[\times [0, 2\pi[\times [0, 2\pi[\mapsto \mathbb{R}$

$$
\begin{aligned}
V(\psi_i, \psi_j, \psi_k) = &-(\gamma_i \cos\psi_{D_i} + \gamma\cos\psi_D)(\cos\psi_i + N_i) \\
&- (\gamma_j \cos\psi_{D_j} + \gamma\cos\psi_D)(\cos\psi_j + N_j) - \gamma\cos\psi_D(\cos\psi_k + N_k) \\
&- \gamma(\cos(\psi_i - \psi_j) + \cos(\psi_i - \psi_k) + \cos(\psi_j - \psi_k) + N_{ij} + N_{ik} + N_{jk})
\end{aligned}
\quad (11.46)
$$

where

$$
N_\alpha = \begin{cases} +1 \text{ if } \overline{\psi}_\alpha = \pi \\ -1 \text{ if } \overline{\psi}_\alpha = 0, \end{cases} \qquad N_{\alpha\beta} = \begin{cases} +1 \text{ if } \overline{\psi}_\alpha - \overline{\psi}_\beta = \pm\pi \\ -1 \text{ if } \overline{\psi}_\alpha - \overline{\psi}_\beta = 0, \end{cases} \quad (11.47)
$$

By definition, $V(\overline{\psi}_i, \overline{\psi}_j, \overline{\psi}_k) = 0$, and it is easy to verify that $V(\psi_i, \psi_j, \psi_k)$ is positive definite for all $(\psi_i, \psi_j, \psi_k) \neq (\overline{\psi}_i, \overline{\psi}_j, \overline{\psi}_k)$, for each equilibrium point in (a), (b) and (c). Finally, computing the partial derivatives of (11.46) we have

$$
\frac{\mathrm{d}V}{\mathrm{d}\tau} = \frac{\partial V}{\partial \psi_i}\frac{\mathrm{d}\psi_i}{\mathrm{d}\tau} + \frac{\partial V}{\partial \psi_j}\frac{\mathrm{d}\psi_j}{\mathrm{d}\tau} + \frac{\partial V}{\partial \psi_k}\frac{\mathrm{d}\psi_k}{\mathrm{d}\tau} = -\left(\frac{\mathrm{d}\psi_i}{\mathrm{d}\tau}\right)^2 - \left(\frac{\mathrm{d}\psi_j}{\mathrm{d}\tau}\right)^2 - \left(\frac{\mathrm{d}\psi_k}{\mathrm{d}\tau}\right)^2 < 0 \quad (11.48)
$$

for all $(\psi_i, \psi_j, \psi_k) \neq (\overline{\psi}_i, \overline{\psi}_j, \overline{\psi}_k)$. Thus (11.46) is a strict Liapunov function and the equilibria are asymptotically stable. □

Theorem 5 (OR gate) *Consider the phase deviation equation (11.44), with $\psi_D = \pi$ and $\gamma_i = \gamma_j > 2\gamma > 0$. Then:*

(a) If $\psi_{D_i} = \psi_{D_j} = 0$, then $(\overline{\psi}_i, \overline{\psi}_j, \overline{\psi}_k) = (0, 0, 0)$ is an asymptotically stable equilibrium point.
(b) If $\psi_{D_i} = 0$, $\psi_{D_j} = \pi$ (respectively $\psi_{D_i} = \pi$, $\psi_{D_j} = 0$), then $(\overline{\psi}_i, \overline{\psi}_j, \overline{\psi}_k) = (0, \pi, \pi)$ (respectively $(\overline{\psi}_i, \overline{\psi}_j, \overline{\psi}_k) = (\pi, 0, \pi)$) is an asymptotically stable equilibrium point.
(c) If $\psi_{D_i} = \psi_{D_j} = \pi$ then $(\overline{\psi}_i, \overline{\psi}_j, \overline{\psi}_k) = (\pi, \pi, \pi)$ is an asymptotically stable equilibrium point.

Proof The proof is completely analogous to theorem 4. □

11.5 Conclusions

In this work we have discussed emerging alternatives to standard computing architectures aiming at overcoming their most challenging limits. This excursus highlighted unconventional approaches in von Neumann architecture configurations as well as in non-von Neumann ones, and it culminated in the discussion of oscillator-based approaches.

We have introduced a formal apparatus able to describe systems of weakly coupled oscillators, at the core of unconventional computing architectures based on autonomous systems. We have shown how to derive the phase equations that we can then use to assess the working principle and the stability of these systems. Moreover, as an example, we have shown how to design asymptotically stable NOT, MAJORITY, AND and OR gates by means of weakly coupled oscillators.

This work provides an important mathematical framework useful for the design and analysis of this emerging computing approach.

References

1. J.J.P. Eckert, J.W. Mauchly, Electronic numerical integrator and computer, US patent # US757158A (1947)
2. H.H. Goldstine, A. Goldstine, The electronic numerical integrator and computer (ENIAC). Math. Tables Other Aids Comput. **2**, 97 (1946)
3. J. von Neumann, First draft of a report on the EDVAC. Ann. History Comput. IEEE **15**(4), 27–75 (1993)
4. J.L. Hennessy, D.A. Patterson, *Computer Architecture, Sixth Edition: A Quantitative Approach* (Morgan Kaufmann Publishers Inc., San Francisco, CA, USA, 2017)
5. S. Arora, B. Barak, *Computational Complexity: A Modern Approach* (Cambridge University Press, Cambridge, 2009)
6. J. Backus, Can programming be liberated from the von Neumann style?: A functional style and its algebra of programs. Commun. ACM **21**, 613–641 (1978). (Aug.)
7. J.L. Hennessy, D.A. Patterson, A new golden age for computer architecture. Commun ACM **62**, 48–60 (2019)
8. M. Horowitz, 1.1 computing's energy problem (and what we can do about it), in *2014 IEEE International Solid-State Circuits Conference Digest of Technical Papers (ISSCC)* (IEEE, 2014)
9. J. Nickolls, W.J. Dally, The GPU computing era. IEEE Micro **30**, 56–69 (2010)
10. D. Singh, C.K. Reddy, A survey on platforms for big data analytics. J. Big Data **2** (2014)
11. Y. LeCun, Y. Bengio, G. Hinton, Deep learning. Nature **521**, 436–444 (2015). (May)
12. I. Goodfellow, J. Bengio, A. Courville, F. Bach, *Deep Learning* (MIT Press Ltd, 2016)
13. V. Sze, Y.-H. Chen, J. Emer, A. Suleiman, Z. Zhang, Hardware for machine learning: Challenges and opportunities, in *2017 IEEE Custom Integrated Circuits Conference (CICC)* (IEEE, 2017)
14. V. Sze, Y.-H. Chen, T.-J. Yang, J.S. Emer, Efficient processing of deep neural networks: A tutorial and survey. Proc. IEEE **105**, 2295–2329 (2017)
15. N.P. Jouppi, C. Young, N. Patil, D. Patterson, G. Agrawal, R. Bajwa, S. Bates, S. Bhatia, N. Boden, A. Borchers, R. Boyle, P. luc Cantin, C. Chao, C. Clark, J. Coriell, M. Daley, M. Dau, J. Dean, B. Gelb, T. V. Ghaemmaghami, R. Gottipati, W. Gulland, R. Hagmann, C. R. Ho, D. Hogberg, J. Hu, R. Hundt, D. Hurt, J. Ibarz, A. Jaffey, A. Jaworski, A. Kaplan, H. Khaitan, D. Killebrew, A. Koch, N. Kumar, S. Lacy, J. Laudon, J. Law, D. Le, C. Leary, Z. Liu, K. Lucke, A. Lundin, G. MacKean, A. Maggiore, M. Mahony, K. Miller, R. Nagarajan, R. Narayanaswami, R. Ni, K. Nix, T. Norrie, M. Omernick, N. Penukonda, A. Phelps, J. Ross, M. Ross, A. Salek, E. Samadiani, C. Severn, G. Sizikov, M. Snelham, J. Souter, D. Steinberg, A. Swing, M. Tan, G. Thorson, B. Tian, H. Toma, E. Tuttle, V. Vasudevan, R. Walter, W. Wang, E. Wilcox, D.H. Yoon, "In-datacenter performance analysis of a tensor processing unit, in *Proceedings of the 44th Annual International Symposium on Computer Architecture* (ACM, 2017)

16. N. Jouppi, C. Young, N. Patil, D. Patterson, Motivation for and evaluation of the first tensor processing unit. IEEE Micro **38**, 10–19 (2018)
17. C. Li, D. Belkin, Y. Li, P. Yan, M. Hu, N. Ge, H. Jiang, E. Montgomery, P. Lin, Z. Wang, W. Song, J.P. Strachan, M. Barnell, Q. Wu, R. S. Williams, J.J. Yang, Q. Xia, Efficient and self-adaptive in-situ learning in multilayer memristor neural networks. Nat. Commun. **9** (2018)
18. Q. Wang, X. Wang, S.H. Lee, F.-H. Meng, W.D. Lu, A deep neural network accelerator based on tiled RRAM architecture, in *2019 IEEE International Electron Devices Meeting (IEDM)* (IEEE, 2019)
19. S.H. Lee, X. Zhu, W.D. Lu, Nanoscale resistive switching devices for memory and computing applications. Nano Res. **13**, 1228–1243 (2020)
20. P. Benioff, The computer as a physical system: A microscopic quantum mechanical Hamiltonian model of computers as represented by turing machines. J. Stat. Phys. **22**, 563–591 (1980)
21. R.P. Feynman, Simulating physics with computers. Int. J. Theor. Phys. **21**, 467–488 (1982)
22. Y. Manin, *Computable and Uncomputable (in Russian)* (Sovetskoye Radio, Moscow, 1980)
23. M.A. Nielsen, I.L. Chuang, *Quantum Computation and Quantum Information* (Cambridge Series on information and the Natural Sciences, Cambridge University Press, 10th Aniversary ed., 2010)
24. P.W. Shor, Algorithms for quantum computation: discrete logarithms and factoring, in *1994 Proceedings of 35th Annual Symposium on Foundations of Computer Science* (1994), pp. 124–134
25. M. Dyakonov, When will useful quantum computers be constructed? Not in the foreseeable future, this physicist argues. here's why: The case against: Quantum computing. IEEE Spectrum **56**, 24–29 (2019). (Mar)
26. S. Yu, Neuro-inspired computing with emerging nonvolatile memorys. Proc. IEEE **106**, 260–285 (2018)
27. G.W. Burr, R.M. Shelby, A. Sebastian, S. Kim, S. Kim, S. Sidler, K. Virwani, M. Ishii, P. Narayanan, A. Fumarola, L.L. Sanches, I. Boybat, M.L. Gallo, K. Moon, J. Woo, H. Hwang, Y. Leblebici, Neuromorphic computing using non-volatile memory. Adv. Phys. X **2**, 89–124 (2016)
28. S.B. Furber, F. Galluppi, S. Temple, L.A. Plana, The SpiNNaker project. Proc. IEEE **102**, 652–665 (2014)
29. M. Davies, N. Srinivasa, T.-H. Lin, G. Chinya, Y. Cao, S.H. Choday, G. Dimou, P. Joshi, N. Imam, S. Jain, Y. Liao, C.-K. Lin, A. Lines, R. Liu, D. Mathaikutty, S. McCoy, A. Paul, J. Tse, G. Venkataramanan, Y.-H. Weng, A. Wild, Y. Yang, H. Wang, Loihi: A neuromorphic manycore processor with on-chip learning. IEEE Micro **38**, 82–99 (2018)
30. I. Boybat, M.L. Gallo, S.R. Nandakumar, T. Moraitis, T. Parnell, T. Tuma, B. Rajendran, Y. Leblebici, A. Sebastian, E. Eleftheriou, Neuromorphic computing with multi-memristive synapses. Nat. Commun. **9** (2018)
31. A. Tavanaei, M. Ghodrati, S.R. Kheradpisheh, T. Masquelier, A. Maida, Deep learning in spiking neural networks. Neural Netw. **111**, 47–63 (2019)
32. G. Singh, L. Chelini, S. Corda, A.J. Awan, S. Stuijk, R. Jordans, H. Corporaal, A.-J. Boonstra, Near-memory computing: Past, present, and future. Microprocessors Microsyst. **71**, 102868 (2019)
33. D. Ielmini, H.-S.P. Wong, In-memory computing with resistive switching devices. Nat. Electron. **1**, 333–343 (2018)
34. F.L. Traversa, F. Bonani, Y.V. Pershin, M. Di Ventra, Dynamic computing random access memory. Nanotechnology **25**, 285201 (2014)
35. Y.V. Pershin, F.L. Traversa, M. Di Ventra, Memcomputing with membrane memcapacitive systems. Nanotechnology **26**, 225201 (2015)
36. A. Sebastian, M.L. Gallo, R. Khaddam-Aljameh, E. Eleftheriou, Memory devices and applications for in-memory computing. Nature Nanotechnol. **15**, 529–544 (2020)
37. F.L. Traversa, M. Di Ventra, Universal memcomputing machines. IEEE Trans. Neural Netw. Learn. Syst. **26**(11), 2702 (2015)

38. M. Di Ventra, F.L. Traversa, Perspective: Memcomputing: Leveraging memory and physics to compute efficiently. J. Appl. Phys. **123**, 180901 (2018)
39. Y.R. Pei, F.L. Traversa, M.D. Ventra, On the universality of memcomputing machines. IEEE Trans. Neural Netw. Learn. Syst. **30**, 1610–1620 (2019)
40. F.L. Traversa, C. Ramella, F. Bonani, M. Di Ventra, Memcomputing NP-complete problems in polynomial time using polynomial resources and collective states. Sci. Adv. **1**(6), e1500031 (2015)
41. F.L. Traversa, M. Di Ventra, Polynomial-time solution of prime factorization and np-complete problems with digital memcomputing machines. Chaos: Interdiscip. J. Nonlinear Sci. **27**, 023107 (2017)
42. F.L. Traversa, M. Di Ventra, Memcomputing integer linear programming (2018). arXiv:1808.09999
43. H. Manukian, F.L. Traversa, M. Di Ventra, "Memcomputing numerical inversion with self-organizing logic gates. IEEE Trans. Neural Netw. Learn. Syst. **99**, 1–6 (2017)
44. M. Di Ventra, F.L. Traversa, Absence of chaos in digital memcomputing machines with solutions. Phys. Lett. A **381**, 3255 (2017)
45. M. Di Ventra, F.L. Traversa, Absence of periodic orbits in digital memcomputing machines with solutions. Chaos: Interdiscipl. J. Nonlinear Sci. **27**, 101101 (2017)
46. F.L. Traversa, P. Cicotti, F. Sheldon, M. Di Ventra, Evidence of exponential speed-up in the solution of hard optimization problems. Complexity **2018**, 1–13 (2018)
47. F. Sheldon, P. Cicotti, F.L. Traversa, M.D. Ventra, Stress-testing memcomputing on hard combinatorial optimization problems. IEEE Trans. Neural Netw. Learn. Syst. **31**(6), 2222–2226 (2020)
48. F.L. Traversa, Aircraft loading optimization: Memcomputing the 5th airbus problem (2019). arXiv:1903.08189
49. F. Sheldon, F.L. Traversa, M.D. Ventra, Taming a nonconvex landscape with dynamical long-range order: Memcomputing ising benchmarks. Phys. Rev. E **100** (2019)
50. H. Manukian, F.L. Traversa, M. Di Ventra, Accelerating deep learning with memcomputing. Neural Netw. **110**, 1–7 (2019)
51. K.Y. Camsari, R. Faria, B.M. Sutton, S. Datta, Stochastic p -bits for invertible logic. Phys. Rev. X **7** (2017)
52. K.Y. Camsari, S. Salahuddin, S. Datta, Implementing p-bits with embedded MTJ. IEEE Electron Device Lett. **38**, 1767–1770 (2017)
53. F. Caravelli, C. Nisoli, Logical gates embedding in artificial spin ice. New J. Phys. **22**, 103052 (2020)
54. E. Goto, New parametron circuit element using nonlinear reactance, in *KDD Kenyku Shiryo* (1954)
55. E. Goto, The parametron, a digital computing element which utilizes parametric oscillation. Proc. IRE **47**, 1304–1316 (1959)
56. J. von Neumann, Non-linear capacitance or inductance switching, amplifying, and memory organs. patent # US2815488A (1954)
57. R. Wigington, A new concept in computing. Proc. IRE **47**, 516–523 (1959)
58. S. Farzeen, G. Ren, C. Chen, An ultra-low power ring oscillator for passive UHF RFID transponders, in *2010 53rd IEEE International Midwest Symposium on Circuits and Systems* (IEEE, 2010)
59. D. Houssameddine, U. Ebels, B. Delaët, B. Rodmacq, I. Firastrau, F. Ponthenier, M. Brunet, C. Thirion, J.-P. Michel, L. Prejbeanu-Buda, M.-C. Cyrille, O. Redon, B. Dieny, Spin-torque oscillator using a perpendicular polarizer and a planar free layer. Nat. Mater. **6**, 447–453 (2007)
60. S. Kobayashi, T. Kimura, Injection locking characteristics of an AlGaAs semiconductor laser. IEEE J. Quantum Electron. **16**, 915–917 (1980)
61. X.L. Feng, C.J. White, A. Hajimiri, M.L. Roukes, A self-sustaining ultrahigh-frequency nanoelectromechanical oscillator. Nat. Nanotechnol. **3**, 342–346 (2008)

62. M.B. Elowitz, S. Leibler, A synthetic oscillatory network of transcriptional regulators. Nature **403**, 335–338 (2000)
63. K.M. Hannay, D.B. Forger, V. Booth, Macroscopic models for networks of coupled biological oscillators. Sci. Adv. **4**, e1701047 (2018)
64. K. Matsuoka, Analysis of a neural oscillator. Biol. Cybern. **104**, 297–304 (2011)
65. F.L. Traversa, Y.V. Pershin, M. Di Ventra, Memory models of adaptive behavior. IEEE Trans. Neural Netw. Learn. Syst. **24**, 1437–1448 (2013)
66. A.T. Winfree, Biological rhythms and the behavior of populations of coupled oscillators. J. Theor. Biol. **16**, 15–42 (1967)
67. J. Roychowdhury, Boolean computation using self-sustaining nonlinear oscillators. Proc. IEEE **103**, 1958–1969 (2015)
68. A. Raychowdhury, A. Parihar, G.H. Smith, V. Narayanan, G. Csaba, M. Jerry, W. Porod, S. Datta, Computing with networks of oscillatory dynamical systems. Proc. IEEE **107**, 73–89 (2019)
69. M. Bonnin, F. Bonani, F.L. Traversa, Logic gates implementation with coupled oscillators, in *2018 IEEE Workshop on Complexity in Engineering (COMPENG)* (IEEE, 2018)
70. G. Csaba, A. Raychowdhury, S. Datta, W. Porod, Computing with coupled oscillators: Theory, devices, and applications, in *2018 IEEE International Symposium on Circuits and Systems (ISCAS)* (IEEE, 2018)
71. T. Wang, J. Roychowdhury, Oscillator-based ising machine (2017). arXiv:1709.08102
72. J. Chou, S. Bramhavar, S. Ghosh, W. Herzog, Analog coupled oscillator based weighted ising machine. Sci. Rep. **9** (2019)
73. A. Mallick, M.K. Bashar, D.S. Truesdell, B.H. Calhoun, S. Joshi, N. Shukla, Using synchronized oscillators to compute the maximum independent set. Nat. Commun. **11** (2020)
74. G. Csaba, W. Porod, Coupled oscillators for computing: A review and perspective. Appl. Phys. Rev. **7**, 011302 (2020)
75. K. Chen, D. Wang, A dynamically coupled neural oscillator network for image segmentation. Neural Netw. **15**, 423–439 (2002)
76. N. Shukla, A. Parihar, M. Cotter, M. Barth, X. Li, N. Chandramoorthy, H. Paik, D.G. Schlom, V. Narayanan, A. Raychowdhury, S. Datta, Pairwise coupled hybrid vanadium dioxide-MOSFET (HVFET) oscillators for non-boolean associative computing, in *2014 IEEE International Electron Devices Meeting* (IEEE, 2014)
77. L.O. Chua, C.A. Desoer, E.S. Kuh, *Linear and Nonlinear Circuits* (McGraw-Hill, New York, 1987)
78. L. Perko, *Differential Equations and Dynamical Systems* (Springer, New York, 2013)
79. R. Albert, A.-L. Barabási, Statistical mechanics of complex networks. Rev. Modern Phys. **74**, 47–97 (2002)
80. S. Boccaletti, V. Latora, Y. Moreno, M. Chavez, D. Hwang, Complex networks: Structure and dynamics. Phys. Rep. **424**, 175–308 (2006)
81. F.C. Hoppensteadt, E.M. Izhikevich, Oscillatory neurocomputers with dynamic connectivity. Phys. Rev. Lett. **82**, 2983–2986 (1999)
82. M. Itoh, L.O. Chua, Star cellular neural networks for associative and dynamic memories. Int. J. Bifurcation Chaos **14**, 1725–1772 (2004)
83. J. Guckenheimer, Isochrons and phaseless sets. J. Math. Biol. **1**, 259–273 (1975)
84. M. Bonnin, F. Corinto, M. Gilli, Phase space decomposition for phase noise and synchronization analysis of planar nonlinear oscillators. IEEE Trans. Circuits Syst. II: Express Briefs **59**, 638–642 (2012)
85. M. Bonnin, F. Corinto, Phase noise and noise induced frequency shift in stochastic nonlinear oscillators. IEEE Trans. Circuits Syst. I: Regular Papers **60**, 2104–2115 (2013)
86. E. Brown, J. Moehlis, P. Holmes, On the phase reduction and response dynamics of neural oscillator populations. Neural Comput. **16**, 673–715 (2004)
87. B. Ermentrout, Type I membranes, phase resetting curves, and synchrony. Neural Comput. **8**, 979–1001 (1996)

88. Y. Kuramoto, *Chemical Oscillations, Waves, and Turbulence* (Springer, Berlin Heidelberg, 2011)
89. H. Nakao, Phase reduction approach to synchronisation of nonlinear oscillators. Contemporary Phys. **57**, 188–214 (2015)
90. J.-N. Teramae, H. Nakao, G.B. Ermentrout, Stochastic phase reduction for a general class of noisy limit cycle oscillators. Phys. Rev. Lett. **102** (2009)
91. K.C. Wedgwood, K.K. Lin, R. Thul, S. Coombes, Phase-amplitude descriptions of neural oscillator models. J. Math. Neurosci. **3**(1), 2 (2013)
92. D. Wilson, J. Moehlis, Isostable reduction of periodic orbits. Phys. Rev. E **94**(5), 052213 (2016)
93. K. Yoshimura, K. Arai, Phase reduction of stochastic limit cycle oscillators. Phys. Rev. Lett. **101** (2008)
94. A. Guillamon, G. Huguet, A computational and geometric approach to phase resetting curves and surfaces. SIAM J. Appl. Dyn. Syst. **8**, 1005–1042 (2009)
95. G. Huguet, R. de la Llave, Computation of limit cycles and their isochrons: Fast algorithms and their convergence. SIAM J. Appl. Dyn. Syst. **12**, 1763–1802 (2013)
96. E. Izhikevich, *Dynamical Systems in Neuroscience: The Geometry of Excitability and Bursting* (MIT Press, Cambridge, Mass, 2007)
97. A. Mauroy and I. Mezić, On the use of Fourier averages to compute the global isochrons of (quasi)periodic dynamics. Chaos: Interdiscip. J. Nonlinear Sci. **22**, 033112 (2012)
98. A. Mauroy, I. Mezić, J. Moehlis, Isostables, isochrons, and Koopman spectrum for the action–angle representation of stable fixed point dynamics. Physica D: Nonlinear Phenomena **261**, 19–30 (2013)
99. D. Wilson, B. Ermentrout, Phase models beyond weak coupling. Phys. Rev. Lett. **123**(16), 164101 (2019)
100. D. Wilson, B. Ermentrout, Augmented phase reduction of (not so) weakly perturbed coupled oscillators. SIAM Rev. **61**(2), 277–315 (2019)
101. D. Aronson, G. Ermentrout, N. Kopell, Amplitude response of coupled oscillators. Physica D: Nonlinear Phenomena **41**, 403–449 (1990)
102. M. Bonnin, F. Corinto, Influence of noise on the phase and amplitude of second-order oscillators. IEEE Trans. Circuits Syst. II: Express Briefs **61**(3), 158–162 (2014)
103. M. Bonnin, F. Corinto, V. Lanza, A mathematical framework for amplitude and phase noise analysis of coupled oscillators. Euro. Phys. J. Special Top. **225**, 171–186 (2016)
104. M. Bonnin, Amplitude and phase dynamics of noisy oscillators. Int. J. Circuit Theory Appl. **45**, 636–659 (2016)
105. M. Bonnin, Phase oscillator model for noisy oscillators. Euro. Phys. J. Special Top. **226**, 3227–3237 (2017)
106. M. Bonnin, F.L. Traversa, F. Bonani, Colored noise in oscillators. phase-amplitude analysis and a method to avoid the Itô-Stratonovich dilemma. IEEE Trans. Circuits Syst. I: Regular Papers **66**, 3917–3927 (2019)
107. A. Demir, A. Mehrotra, J. Roychowdhury, Noise in oscillators: A review of state space decomposition approaches. IEEE Trans. Circuits Syst. I: Fund. Theory Appl. **47**, 655–674 (2000)
108. F.X. Kaertner, Analysis of white and $f^{-\alpha}$ noise in oscillators. Int. J. Circuit Theory Appl. **18**, 485–519 (1990)
109. F.L. Traversa, M. Bonnin, F. Corinto, F. Bonani, Noise in oscillators: A review of state space decomposition approaches. J. Comput. Electroni. **14**, 51–61 (2014)
110. D. Wilson, Phase-amplitude reduction far beyond the weakly perturbed paradigm. Phys. Rev. E **101**(2), 022220 (2020)
111. D. Wilson, A data-driven phase and isostable reduced modeling framework for oscillatory dynamical systems. Chaos: Interdiscip. J. Nonlinear Sci. **30**(1), 013121 (2020)
112. T. Djurhuus, V. Krozer, J. Vidkjær, T.K. Johansen, Oscillator phase noise: A geometrical approach. IEEE Trans. Circuits Syst. I: Regular Papers **56**(7), 1373–1382 (2008)
113. T. Djurhuus, V. Krozer, A study of amplitude-to-phase noise conversion in planar oscillators. Int. J. Circuit Theory Appl. (2020)

114. F.L. Traversa, F. Bonani, Frequency-domain evaluation of the adjoint Floquet eigenvectors for oscillator noise characterisation. IET Circuits, Dev. Syst. **5**(1), 46 (2011)
115. F.L. Traversa, F. Bonani, Improved harmonic balance implementation of Floquet analysis for nonlinear circuit simulation. AEU - Int. J. Electron. Commun. **66**, 357–363 (2012)
116. F.L. Traversa, F. Bonani, Selective determination of Floquet quantities for the efficient assessment of limit cycle stability and oscillator noise. IEEE Trans. Comput.-Aided Des. Integr. Circuits Syst. **32**, 313–317 (2013)
117. W. Wernick, Complete sets of logical functions. Trans. Am. Math. Soc. **51**(1), 117–132 (1942)

Chapter 12
Design and Implementation in a New Approach of Non-minimal State Space Representation of a MIMO Model Predictive Control Strategy-Case Study and Performance Analysis

Roxana-Elena Tudoroiu, Mohammed Zaheeruddin, Sorin Mihai Radu, Dumitru Dan Burdescu, and Nicolae Tudoroiu

Abstract The paper's main objective is to investigate a new design approach of a model predictive control strategy in closed-loop applied to minimal and nonminimal state-space realizations implemented in real-time on a MATLAB SIMULINK platform. The proposed control strategy's effectiveness is illustrated on a centrifugal chiller system, chosen as a case study. The new control design lays on the linear discrete-time multi-input, multi-output autoregressive moving average models, obtained by following the system identification procedure provided in MATLAB Identification Toolbox. The reason is to simplify the high complexity of centrifugal chiller dynamics, in terms of dimension and encountered nonlinearities; thus, the models become more practical and straightforward for real-time implementation of the proposed control strategy. They are also handy for closed-loop simulation purpose, and useful to prove the effectiveness of the proposed model predictive closed-loop control strategy applied to the case study. The MATLAB SIMULINK simulation results show a fast-transient speed, high tracking accuracy, and an excellent robustness to possible changes in the load disturbance and sensor measurements

R.-E. Tudoroiu
Mathematics and Informatics, University of Petrosani, Petrosani, Romania
e-mail: tudelena@mail.com

M. Zaheeruddin
Building Civil and Environmental Engineering, Concordia University, Montreal, Canada
e-mail: zaheer@encs.concordia.ca

S. M. Radu
Mechanical and Electrical Engineering, University of Petrosani, Petrosani, Romania

D. D. Burdescu
Software Engineering Department, University of Craiova, Craiova, Romania

N. Tudoroiu (✉)
Engineering Technologies Department, John Abbott College, Sainte-Anne-de-Bellevue, Canada

M. Virvou et al. (eds.), *Advances in Artificial Intelligence-based Technologies*, Learning and Analytics in Intelligent Systems 22,
https://doi.org/10.1007/978-3-030-80571-5_12

noise levels. These results are encouraging for the new developments and extensions in the future work.

Keywords Centrifugal chiller · System identification and ARMAX models · State-space representation · MPC · MPC Designer Simulink block · PI controller

12.1 Introduction

The centrifugal chiller system is one of the extensively used subsystems in a variety of heating, ventilation, and air conditioning (HVAC) systems applications, due to its "...high capacity, high reliability, and low maintenance", as is shown in [1]. The primary role played by a centrifugal chiller integrated within an HVAC system architecture used for control purpose of the leading environmental parameters inside the buildings such as temperatures, pressures, flows, and humidity is to provide through two water loops the chilled-water required to cover the cooling demand of all air handling units. One of the chiller loops is acted by a water pump to circulate the chilled water through the evaporator and throughout the building for cooling, and a second water pump acts a second water loop. The heat is rejected from the water inside the condenser by circulating the heated water to the cooling tower and cooled back. Further details on technical aspects of the system are well documented in the references [1–6]. The efficiency of the centrifugal chiller must be improved since it is one of the most energy-consuming devices. An intensive literature review in HVAC control systems field reveals a significant amount of research done by Li Pengfei et al. in [2, 5]. This research lays on transient and steady-state modelling for centrifugal chillers.

Regarding the accuracy of the modelling procedure, it is worth noting that in [2] the system of differential–algebraic equations describing the centrifugal chiller dynamics is consistently initialized. In 2011 Guang-Yu Jin et al. has proposed in [3] the concept of hybrid model for water-cooled centrifugal chiller, for which the modelling procedure "....takes advantages of both physical and empirical modelling approaches". It can also "predict the centrifugal chiller performance "accuratelyin wide operating range and in real-time", significantly reducing the demand for computational load growth. The model is validated based on the experimental data set collected from a testing system. Moreover, the testing results show that the centrifugal chiller proposed model "...can predict the performance of the real-time operating chiller accurately" and can be used successfully in a wide range of applications such as operational optimization, performance assessment, and to develop several fault detection, diagnosis, and isolation (FDDI) strategies. The reduction of today CO2 emissions becomes a critical consideration in the technical improvement and product development of the centrifugal chiller equipment. HVAC heat sources latest trend and the new concept of "eco Turbo ETI" series of high-efficiency centrifugal chillers have inverters as standard equipment. Thus, new concepts for user-friendly centrifugal chiller equipment are fully justified since it is easy to install. There is a growing demand in a wide range of HVAC applications of high-efficiency centrifugal chiller systems. In [4] is presented a new method for the optimal chiller sequencing control (CSC) using a model predictive control (MPC) strategy. The authors have proposed for a multi-chiller plant a new cooling load-based chiller sequencing control (CLC) method implemented in a Model Predictive Control (MPC) framework. The optimization objective is "...the minimization of the chiller plant's total energy consumption, including chillers, cooling towers and pumps. It is performed continuously according to the predicted cooling load and wet bulb temperature" [4].

According to the same document [4] "...the independent variables are the thresholds for chiller staging and the condenser water set point". Recent advances in dynamic modelling of HVAC equipment are presented in [5], especially using Modelica that offers two basic core features, i.e., object-oriented modelling platform and a causality. It is best suited to address HVAC equipment challenges, and significantly reduces the modelling efforts and time-consuming model development. The document also reveals the current issues and the possible future directions of dynamic modelling for HVAC equipment. In [6] a multivariate polynomial empirical novel regression model is proposed for evaluating the coefficient of performance (COP) of centrifugal chillers, cost-effectively and efficiently to be used for on-filed engineers. It is validated and analyzed with both internal and external on-filed measurements data set. The model variables used to describe its dynamic behaviour are the standard parameters on the chilled water side in chiller systems, namely pressure and temperatures. As a novelty, it is worth noting that the model is independent of the cooling load or the chilled water flow rate, so difficult to be measured directly and accurately in many on-fields operating chiller plants. An interesting idea of systemic decomposition of the centrifugal chiller plant in two independent control loops is presented and implemented [7]. More precisely, the research done in [7] considers the centrifugal chiller as an interconnection of two main multi-input multi-output (MIMO) embedded control loops subsystems, such as the chilled water temperature inside the evaporator, and a refrigerant liquid control level inside the condenser. In this paper, the preliminary results obtained in [7] related to the real-time implementation are extended in a new MATLAB model-based platform of a single-input single-output (SISO) predictive proportional integral-plus (PIP) control strategies [9–15]. The centrifugal chiller plant is described in a nonminimal state space (NMSS) representation [10–15], by using a discrete-time SISO autoregressive moving average with exogenous input (ARMAX) models. The ARMAX models are generated by following an identification procedure and extensive simulations in MATLAB R2017b [16]. Their accuracy is improved in [8] through a combination of neural networks and a fuzzy logic approach. Besides, in [7], Appendix 1, pp. 299–305, is presented the complete and most accurate centrifugal chiller model that is of great complexity and highly nonlinear. Extensive open-loop simulations in real-time MATLAB environments using an accurate centrifugal chiller model are carried out for the new control design approach to get the input–output measurements dataset for identification and parameters estimation of linear discrete-time MIMO ARMAX models [16]. Based on these models in the new design approach are developed in both nonminimal state space (NMSS) and minimal state space (MSS) realizations two predictive MIMO MPC closed-loop control strategies [15], using the MATLAB SIMULINK MPC Designer [17], MPC online Toolbox [18], and a beneficial MPC User's Guide [19]. The design and real-time MATLAB implementation of MPC strategy for minimal state space realization follows the procedure suggested in [18–20]. A rigorous analysis of the centrifugal chiller closed-loop control performance reveals a great tracking accuracy, robustness to step input changes, fast transient speed, reduced overshoot amplitude, and disturbance rejection. The performance analysis of both MPC control strategies provides valuable information to select the most suitable and accurate control strategy that performs better, meets the control design requirements and process control constraints. It must also reject the effect of possible disturbances that act on the controlled process. The remainder of this paper is organized as follows. Section 12.2 is focused on systemic decomposition of the centrifugal chiller system and description of its dynamic MIMO ARMAX model. Sections 12.3 and 12.4 propose two MIMO MPC strategies for minimal and nonminimal state-space centrifugal chiller system realizations whose design lays on MATLAB SIMULINK 2018 MPC Designer block. Section 5 concludes the main contributions of the paper.

12.2 Centrifugal Chiller—System Decomposition

A simplified schematic of the proposed centrifugal chiller under investigation in our case study is shown in Fig. 12.1.

This schematic of centrifugal chiller involves the following four main components: (1) centrifugal compressor; (2) condenser; (3) expansion valve; (4) evaporator; details are documented in [7]. For system decomposition purpose the chilled water plant can be split in one circuiting refrigerant loop and two water loops. The first water loop is circuited between the condenser and the cooling tower, and the second water loop is circuited between the evaporator and the air handling units (AHU) that produces chilled water for the cooling coil, as is also mentioned in [7].

12.2.1 Centrifugal Chiller Dynamic Model Description

As is mentioned in previous section, the open-loop simulations of the transient and steady-state behaviour of the water-cooled centrifugal chiller in this research is performed in a real timeon MATLAB/SIMULINK platform based on the full model introduced in [23], Appendix 1, pp. 299–305. It is worth mentioning two main features for centrifugal chiller dynamic model proposed to be investigated in [23], namely a great complexity in terms of state dimension and its high nonlinearity. In addition, the model is described in state-space representation by a set of first-order

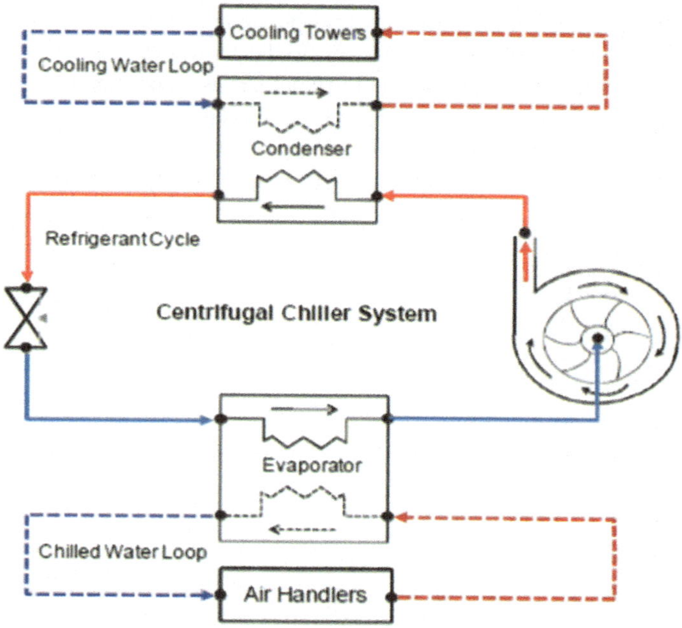

Fig. 12.1 Schematic of water-cooled centrifugal chiller system (see [2])

differential equations (ODE), in a mathematically appropriate form to represent a physical system.

12.2.2 Centrifugal Chiller Dynamic MIMO ARMAX Model Description

In this section is presented alinear MIMO ARMAX chiller plant model with three inputs and two outputs that for simplicity might be also decoupled into two multi input and one single output (MISO) models, first one attached to Evaporator and the second attached to Condenser, the two embedded subsystems of the centrifugal chiller control plant.Based on MIMO ARMAX model proposed in this section, two closed loop MIMO MPC strategies are built and implemented in real-time. Thelinear MIMO ARMAX model is described in discrete timegenerated on MATLAB R2018b platform through specific subroutines found inSystem Identification Toolbox, as is shown in [16].The MIMO ARMAX model parameters are estimated through MATLAB "armax" function based on the inputoutput measurement dataset provided by an extensive amount of open loop MATLAB simulations performed on the centrifugal chiller plant. Theoretically, the MATLAB function "armax" is a subroutine based on least squared estimation (LSE) method and on a selection procedure of optimal values for polynomial orders. Additionally, pure transport delays in the signals flowing in the feedback path from the measurement sensors to the controllers for each input/output pair are also specified in "armax" routine structure, as is mentioned in [7, 16]. Moreover, in each MIMO ARMAX centrifugal chiller model are incorporated "coloured" or "white" noise channels (see also [7, 16]). The general polynomial form of a discrete time MIMO ARMAX model looks to one given in [7, 9, 10, 16]:

$$\begin{aligned} A(q)y(t) &= B(q)u(t-n_k) + C(q)e(t) \\ A(q) &= 1 + A_1q^{-1} + A_2q^{-2} + \cdots + A_{n_a}q^{-n_a} \\ B(q) &= B_1 + B_2q^{-1} + B_3q^{-2} + \cdots + B_{n_b}q^{-n_b+1} \\ C(q) &= 1 + C_1q^{-1} + C_2q^{-2} + \cdots + C_{n_c}q^{-n_c} \end{aligned} \tag{12.1}$$

that is equivalent in complex domain {z} to the following description:

$$\begin{aligned} A(z^{-1})Y(z) &= B(z^{-1})U(z)z^{-n_k} + C(z^{-1})E(z) \\ A(z^{-1}) &= 1 + A_1z^{-1} + A_2z^{-2} + \cdots + A_{n_a}z^{-n_a} \\ B(z^{-1}) &= B_1 + B_2z^{-1} + B_3z^{-2} + \cdots + B_{n_b}z^{-n_b+1} \\ C(z^{-1}) &= 1 + C_1z^{-1} + C_2z^{-2} + \cdots + C_{n_c}z^{-n_c} \end{aligned} \tag{12.2}$$

obtained straightforwardly by replacing formally the discrete-time forward shift operator q by a complex variable z. All the variables appearing in Eqs. (12.1) and (12.2) have the same meaning as those defined in [7], and for a better understanding of the document content we have found useful to be repeated:

- $y(t)$ is the output vector at discrete instant t
- n_a, n_b, n_c represent the degrees of the polynomial matrices $A(q)$, $B(q)$, and $C(q)$ respectively
- $u(t)$ is the plant input vector at the discrete instant t
- n_k is an integer number of sampling periods, a pure transport delay of the signal flow between the measurement sensors and controllers in the feedback path.
- A_i, B_i, C_i represent the coefficients (parameters) matrices of the polynomial matrices $A(q)$, $B(q)$, and $C(q)$ of ARMAX models respectively,
- q forward shift operator, i.e. $q(y(t)) = y(t+1)$, and q^{-1} backward shift operator, i.e. $q^{-1}(y(t)) = y(t-1)$, $t = \frac{t}{T_s}$, denotes the normalized sampling time, where $t = kT_s, k \in^+$ is used to designate the description of dynamical discrete-time system, T_s representing the sampling period
- $e(t)$ denotes the "white" noise disturbance vector at the discrete instant t.

- $z \in C, z = e^{st}, s = \sigma + j\omega, \sigma = \mathrm{Re}(s)$, and $\omega = \mathrm{Im}(s)$, representing the real and imaginary parts of the complex variable s respectively.
- $U(z) = Z\{u(t)\}, Y(z) = Z\{y(t)\}, E(z) = Z\{e(t)\}$ denotes the z-images of the input $u(t)$, output $y(t)$, and of the "white" noise $e(t)$. Z denotes also the "transform Z" operator that is applied tothe discrete - time variables $u(t)$, $y(t)$, and $e(t)$.

12.2.3 Centrifugal Chiller Open Loop MIMO ARMAX Discrete-Time Model

If the load disturbance is considered as a new deterministic input the coefficients of the polynomial matrices $A(z^{-1})$, $B_1(z^{-1})$, $B_2(z^{-1})$, $C(z^{-1})$ of the MIMO ARMAX centrifugal chiller plant model, appearing also in the generic Eqs. (12.1) and (12.2), are estimated by:

$$\begin{aligned} A(z^{-1}) &= I_2 + \begin{bmatrix} -1.756 & 0 \\ 0 & -1.626 \end{bmatrix} z^{-1} + \begin{bmatrix} 0.7611 & 0 \\ 0 & 0.6884 \end{bmatrix} z^{-2} \\ B(z^{-1}) &= \begin{bmatrix} -0.02809 & -0.008112 & 0.001175 \\ 1.776 & -2.992 & 0.03049 \end{bmatrix} z^{-1} \\ &+ \begin{bmatrix} 0 & 0.001722 & 0 \\ 0 & 2.071 & 0 \end{bmatrix} z^{-2} \\ C(z^{-1}) &= I_2 + \begin{bmatrix} 0.2993 & 0 \\ 0 & 0.7133 \end{bmatrix} z^{-1} \end{aligned} \tag{12.3}$$

Sample time: 1 s.

Parameterization: the orders of polynomials are set to $n_a = 2$, $n_b = [1\ 2\ 1]$, $n_c = 1$, $nk = [1\ 1\ 1]$ that fit the estimation data: 99.88% (prediction focus), MSE: 1.995e-08 for chilled water temperature in Evaporator open loop, and 87.63% (prediction focus), and FPE: 0.01579, MSE: 0.01563 for refrigerant liquid level in Condenser open loop, respectively.

This description is equivalent to following four state variables MIMO ARMAX centrifugal chiller model state-space representation:

$$\begin{aligned} x_{t+1} &= Ax_t + Bu_t = Ax_t + B_1 u_{1_t} + B_2 u_{2t} \\ y_t &= Cx_t + Du_t \end{aligned} \tag{12.4}$$

where the matrices A, B, C, D are given by:

$$A = \begin{bmatrix} 0 & -0.7611 & 0 & 0 \\ 1 & 1.756 & 0 & 0 \\ 0 & 0 & 0 & -0.6884 \\ 0 & 0 & 1 & 1.626 \end{bmatrix}$$

$$B = \begin{bmatrix} 0 & 0.006886 & 0 \\ -0.1124 & -0.03245 & 0.004177 \\ 0 & 1.035 & 0 \\ 0.8881 & -1.496 & 0.01355 \end{bmatrix} = \begin{bmatrix} B_1 & B_2 \end{bmatrix}$$

$$B_1 = \begin{bmatrix} 0 & 0.006886 \\ -0.1124 & -0.03245 \\ 0 & 1.035 \\ 0.8881 & -1.496 \end{bmatrix}, B_2 = \begin{bmatrix} 0 \\ 0.004177 \\ 0 \\ 0.01355 \end{bmatrix}$$

$$C = \begin{bmatrix} 0 & 0.25 & 0 & 0 \\ 0 & 0 & 0 & 2 \end{bmatrix}$$

$$D = O_{2\times 3}$$

$$x_t = \begin{bmatrix} x_{1,t} \\ x_{2,t} \\ x_{3,t} \\ x_{4,t} \end{bmatrix}, u_t = \begin{bmatrix} u_{1,t} \\ u_{2,t} \\ u_{3,t} \end{bmatrix} = \begin{bmatrix} \text{compressor speed} \\ \text{expansion valve opening} \\ \text{load disturbance} \end{bmatrix} = \begin{bmatrix} u_{com} \\ u_{EXP} \\ T_{rr} \end{bmatrix}$$

$$y_t = \begin{bmatrix} y_{1,t} \\ y_{2,t} \end{bmatrix} = \begin{bmatrix} Tchw \\ level \end{bmatrix}, u_{1t} = \begin{bmatrix} \text{compressor speed} \\ \text{expansion valve opening} \end{bmatrix} = \begin{bmatrix} u_{com} \\ u_{EXP} \end{bmatrix}$$

$$u_{1t} = [\text{load disturbance}] = [T_{rr}]$$

Discretization sampling time is setup to $T_s = 1[s]$.The matrix B is split in two matrices B_1 and B_2 for simulation purpose, suitable to identify the manipulated variables (MV)$u_{1t} = \begin{bmatrix} u_{com} \\ u_{EXP} \end{bmatrix}$ and the measured disturbance $u_{1t} = [T_{rr}]$,useful to formulate the optimization problem (18), to setup and tune the parameters of SIMULINK model and MPC Designershown in Figs. 12.3 and 12.4, respectively.

In Fig. 12.2 is gathered a valuable information about the validation of both SISO ARMAX models, namely for chilled water temperature loop in Evaporator, and for refrigerant liquid level loop in Condenser with three inputs for entire range of the input–output measurements data set, i.e. 3607 samples:

To validate these two SISO ARMAX models MATLAB is used a typical function called "*compare*" found in Identification MATLAB Toolbox [7, 16]. For a MIMO ARMAX centrifugal chiller model, the accuracy of the curves in Fig. 12.2 increases considerable, especially for the refrigerant liquid level.

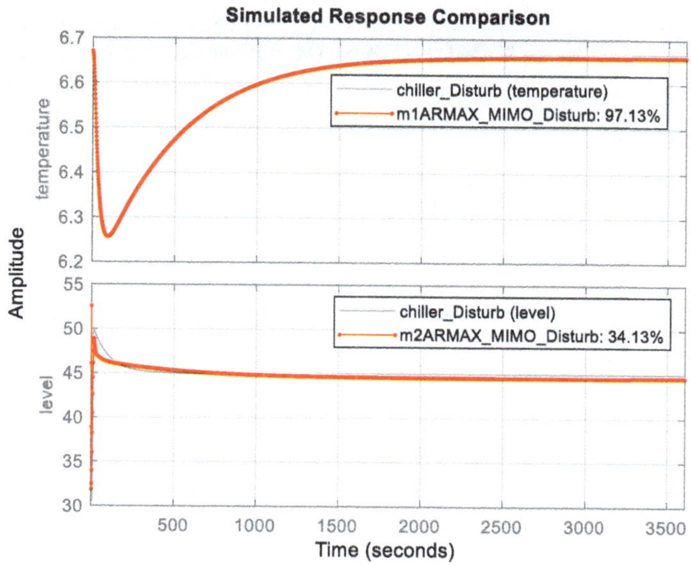

Fig. 12.2 The validation of the estimated ARMAX MIMO models with three inputs (including the load disturbance)

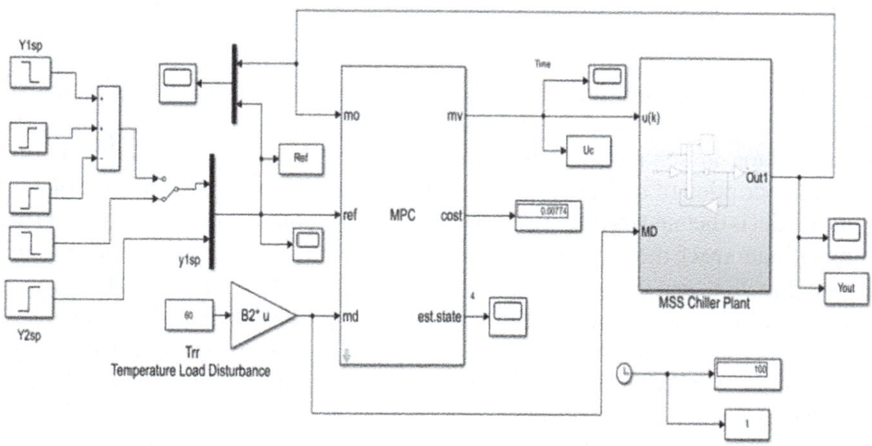

Fig. 12.3 SIMULINK model of the closed loop centrifugal chiller control system with MPC used in MATLAB simulations

12.2.4 Centrifugal Chiller Dynamic MIMO ARMAX Model Nonminimal State Space Description

The novelty of this subsection is a new design control approach by introducing the concept of non-minimal state space (NMSS) control system, as is formulated for

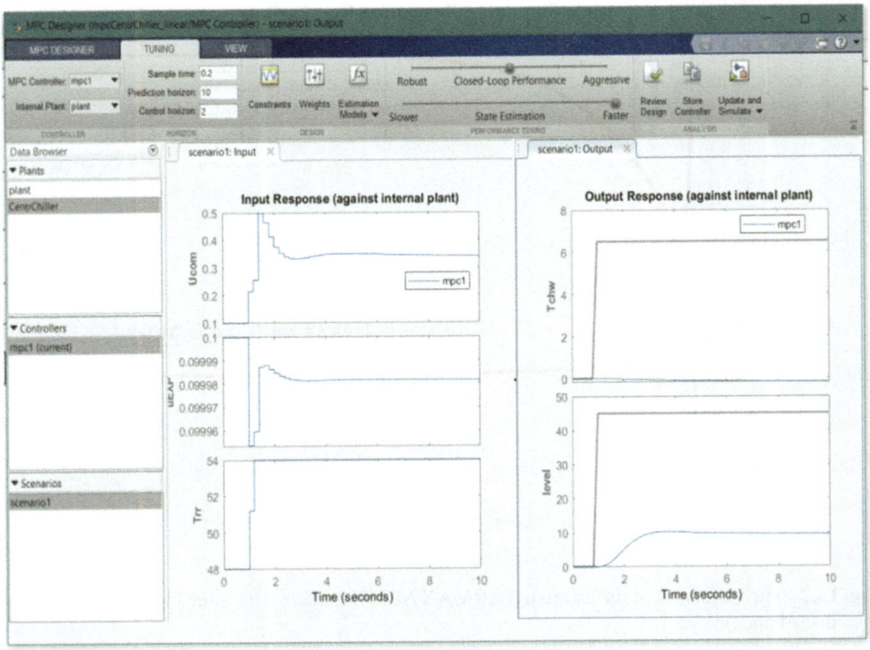

Fig. 12.4 MPC Designer SIMULINK block for control design and parameters tuning

SISO ARX or ARMAX models in [7, 10–14], and for MIMO ARMAX models in [11, 15]. According to these documents, the non-minimal state space (NMSS) representation in contrast to minimal state-space realizations "...seems to be the natural description of a discrete-time transfer function, since the complete structure of the model is dictated by its dimension". The same documents state that "...from the minimal state space descriptions account only for the order of the denominator of the transfer function, and the state variables assigned to each description not always have physical meaning, usually representing combinations of input and output signals".

The proposed MIMO MPC control algorithm in the new approach s known in the literature as MIMO MPC within the NMSS context. It "can be interpreted as a logical extension of the conventional MIMO MPC controller for minimal state-space representation, facilitating its straightforward implementation using a standard hardware-software arrangement". In the paper, the new non-minimal state-space development of the proposed MIMO MPC controller design, follows the same methodologies described in [13, 15], being encouraged by their excellent results for MIMO ALSTOM nonlinear gasifier control plant [13] and many other applications developed in [15].

In NMSS design approach, the MIMO MPC has more design flexibility and robustness than the minimal state space (MSS) representation approach, and shares some features found to a predictive Proportional Integral-Plus (PIP) control design. These aspects are documented and well described in [23], [26–31], within the same

NMSS context. The MIMO ARMAX models in an MSS description given in (12.1)–(12.4) can be represented equivalently in an NMSS representation by the following equations:

$$x(k) = Fx(k-1) + Gu(k-1) + Dy_{sp}(k) \tag{12.5}$$

$$y(k) = Hx(k) \tag{12.6}$$

where the $n + m + n_k - 1$ dimensional NMSS state vector $x(k)$ is defined as,

$$x(k) = [y(k)\ \ y(k-1)...y(k-n-1)\ \ u(k-1)\ u(k-2)...u(k-m-n_k+2)\ \ z(k)]^T \tag{12.7}$$

The new variable $z(k)$ introduced in (12.7) represents in discrete time the integral of error between the set point input (reference) $y_{sp}(k)$ and the control system output $y(k)$:

$$z(k) = z(k-1) + (y_{sp}(k) - y(k)) \tag{12.8}$$

The matrix F, and the vectors g, d, h are calculated based on the methodology described in [29] for a pure delay $n_k = 2$ and the number of inputs equals the number of outputs, and also for $n_k \geq 2$. For details, one can see the general description used for MIMO systems in [29] with the same number of inputs and outputs, and in [31], the number of inputs and outputs might be different. Unlike the minimal state-space realization of MIMO ARMAX centrifugal model given by (12.4) within the NMSS approach the proposed model has (a) five inputs, three of them considered as manipulated variables (MV),

$$u_t = \begin{bmatrix} u_{com} \\ u_{EXP} \\ T_{rr} \end{bmatrix}$$

and the last two inputs:

$$y_{sp}(k) = \begin{bmatrix} y_{1sp} \\ y_{2sp} \end{bmatrix} = \begin{bmatrix} T_{chw\_sp} \\ level\_sp \end{bmatrix}$$

representing the reference setpoints for chilled water temperature and the refrigerant liquid level, involved in the third term of state Eq. (12.5) $Dy_{sp}(k)$ is considered as measured disturbance (MD), required for MPC design and tuning, (b) 9 state variables, given in (12.7), and (c) two outputs:

$$y(k) = \begin{bmatrix} y_1(k) \\ y_2(k) \end{bmatrix} = \begin{bmatrix} T_{chw}(k) \\ level(k) \end{bmatrix}.$$

The proposed MIMO MPC control design follows the procedure described in [31, 33–35]. According to [31], the matrices *F, G, D, H* that appear in (12.5) and (12.6) are straightforwardly calculated based on the set of Eq. (12.3), as follows:

$$\mathrm{F} = \begin{bmatrix} -A_1 & -A_2 & B_2 & Z_2 \\ I_2 & Z_2 & Z_{23} & Z_2 \\ Z_{32} & Z_{32} & Z_{32} & Z_3 \\ A_1 & A_2 & -B_2 & I_2 \end{bmatrix}, \tag{12.9}$$

$$\mathrm{G} = \begin{bmatrix} B_1^T & Z_{32} & I_3 & -B_1^T \end{bmatrix}^T \tag{12.10}$$

$$D = \begin{bmatrix} Z_2 & Z_2 & Z_2 & I_2 \end{bmatrix}^T \tag{12.11}$$

$$H = \begin{bmatrix} I_2 & Z_2 & Z_2 & Z_{23} \end{bmatrix}^T \tag{12.12}$$

$$Z_2 = O_{2\times 2},\ Z_{23} = O_{2\times 3},\ Z_{32} = O_{3\times 2} \tag{12.13}$$

$$I_2 = \begin{bmatrix} 1 & 0 \\ 0 & 1 \end{bmatrix},\ I_3 = \begin{bmatrix} 1 & 0 & 0 \\ 0 & 1 & 0 \\ 0 & 0 & 1 \end{bmatrix} \tag{12.14}$$

where

$$A_1 = \begin{bmatrix} -1.756 & 0 \\ 0 & -1.626 \end{bmatrix},\ A_2 = \begin{bmatrix} 0.7611 & 0 \\ 0 & 0.6884 \end{bmatrix} \tag{12.15}$$

$$B_1 = \begin{bmatrix} -0.02809 & -0.008112 & 0.001175 \\ 1.776 & -2.992 & 0.03049 \end{bmatrix},\ B_2 = \begin{bmatrix} 0 & 0.001722 & 0 \\ 0 & 2.071 & 0 \end{bmatrix}$$

According to (12.7), the structure of the new state vector is given by:

$$x(k) = \begin{bmatrix} y(k) & y(k-1) & u(k-1) & z(k) \end{bmatrix}^T \tag{12.16}$$

for the following structures of the input, output and the integral of error vectors:

$$\begin{aligned} y(k) &= \begin{bmatrix} y_1(k) & y_2(k) \end{bmatrix},\ u(k) = \begin{bmatrix} u_1(k) & u_2(k) & u_3(k) \end{bmatrix} \\ z(k) &= \begin{bmatrix} z_1(k) & z_2(k) \end{bmatrix} \end{aligned} \tag{12.17}$$

It is noteworthy that the state's vector size in the NMSS description given by (12.16) increases from 4 to 9 compared to its size in a minimum state space such as that given in (12.4). Also, the vector element of the vector G occupies the same position as the one occupied by the state vector defined by (12.16). The other components of the vectors G, D, H are easily identified due to the simplicity of their structures.

12.3 MISO MPC Strategy Design in a Minimal State Space Realization

This section presents how can be designed a Model Predictive Control (MPC) strategy in the context of proposed case study, of low computational cost and practical centrifugal chiller control system. Thus, we attempt to solve the centrifugal chiller plant's control by using one MPC strategy to handle constraints imposed on both actuators explicitly.

We are also determined to considerably reduce the computational implementation load for the practical use of the MPC based on a fourth-order prediction MIMO ARMAX model.

The proposed MPC strategy requires a centrifugal chiller model to predict the chiller plant's future along a finite time horizon, as is also mentioned in [36]. In this research paper we use the centrifugal chiller plant MIMO ARMAX linear model in a discrete-time minimal state space (MSS) representation given by matrix Eq. (12.4), from which is removed unmeasurable disturbance term as a prediction model. However, a simple fourth-dimensional predictive model (i.e., only four state variables) described in the previous Sect. 12.2.2 is strongly desired to reduce the microprocessor's computation cost significantly. In the majority of building systems' structures is integrated a Building Energy Management System (BEMS). It predicts the energy performance and optimizes the operations on a various "transient objects which interact with many nonlinear building systems, such as HVAC, chillers, boilers, pumps, fans, and lights", as is mentioned in [21]. Due to the transient behaviour and nonlinear nature of chiller plants, simple controllers such as on/off control, different Proportional Derivative Integral (PID) controller structures, are not enough to predict future states of the chiller plant closed-loop control systems [21]. Overall, an MPC strategy involves four main requirements: (a) a system model; (b) a cost function; (c) constraints; and (d) an optimization algorithm which "can be used for anticipatory performance prediction and control actions over the time horizon", as can also be found in [15, 21–24]. Moreover, in [21] is stated that for a reliable MPC real-time implementation, "the system model must be carefully chosen, related to following features: (a) simplified energy calculation; (b) dynamic simulation; (c) lumped simulation model; and (d) data-driven inverse model".

12.3.1 MIMO MPC Optimization Problem Formulation

The proposed MIMO MPC control design, optimization problem formulationand the notations are like those used in [15, 18–24].

The optimal value of control input vector u has three components (i.e.,compressor speed, expansion valve opening command and chilled water temperature load disturbance), is calculated by solving the generic constrained optimization problem below during each sampling period under different forms such in [20–24]:

$$\min_{u \in U} J = \sum_{k=1}^{p} \{y_{t+k}^T Q_t y_{t+k} + \Delta u_{t+k}^T R_t^{\Delta u} \Delta u_{t+k} + u_{t+k}^T R_t^u u_{t+k}\} \tag{12.18}$$

subjected to:

$$\begin{aligned} x_{t+1} &= Ax_t + Bu_t \\ y_t &= Cx_t + Du_t \\ y_t^{\min} &\le y_t \le y_t^{\max} \\ \Delta u_t^{\min} &\le \Delta u_{t+k} \le \Delta u_t^{\max} \\ u_t^{\min} &\le u_{t+k} \le u_t^{\max} \end{aligned} \tag{12.19}$$

where t is the current time, p is the prediction horizon defined as number of integer intervals, Δu_* is the increment of the control input, and the suffix "min" and "max" in the inequality constraints denotes lower bound and upper bound. Q_t, $R_t^{\Delta u}$ and R_t^u are the positive-semi definite weight matrices for the following error, the change rate and the magnitude of the control input, respectively.

The linear MPC optimization problem above results in a Quadratic Programming (QP) problem. Hence, QP solvers are available such as active-set methods and interior-point methods to solve the problem. Once an optimal solution, i.e., a control input sequence [36]:

$$u^* = [u_t^* \, u_{t+1}^* \, u_{t+2}^* \dots u_{t+p-1}^*]^T \tag{12.20}$$

along prediction horizon is numerically obtained, is consideredonly the first element of the sequence u_t^* as an actual control input. Forthe centrifugal chiller optimization problem, the chilled water temperature y_1, and the refrigerant liquid level y_2 must be regulated to converge to their set points under the upper and lower limits of the inputvector u and outputvector y components. Also, in [36] for the weight matrices Q_t, $R_t^{\Delta u}$ and R_t^u in (12.19) are chosen of thefollowing form:

$$Q_t = \begin{bmatrix} w_{y_1}(t) & 0 \\ 0 & w_{y_2}(t) \end{bmatrix}, R_t^{\Delta u} = w_{\Delta u}(t), R_t^u = w_u(t) \tag{12.21}$$

It is worth noting that each weight value in (12.21) can be changed at every sampling time.

12.3.2 MIMO MPC Parameters Design

To achieve a satisfactory centrifugal performance for chiller control system, is required for this that the parameters of the MIMO MPC controller to be meticulously designed and tuned [20]. The following main parameters must be designed and tuned for a MIMO MPC strategy: (a) the prediction horizon (p); (b) the control horizon (m); and (c) the weights in the performance index (12.18). This subsection is showing the parameter design and tuning strategy obtained through several closed-loop simulation and experiment. MATLAB Simulink and Model Predictive Control Toolbox™ well documented in [18, 19] are used in all simulations. The optimization problem is solved using a QP solver, integrated in the MPC Designer App SIMULINK block. This block is provided by Model Predictive Control Toolbox [19] to model the controller and run closed-loop simulations with the plant model in Simulink, as is shown in Figs. 12.3, and 12.4.

The control design, tuning parameters, input–output constraints (hard and soft) and weights, together with the simulation options are set to the following optimal values of the performance index (J):

a. Ts = 0.2 s (sampling time), p = 10 s (prediction horizon), m = 2 s (control horizon)
b. Input nominal values: U = [0.1 0.1 48]
c. Output nominal values: Y = [0 0]
d. Inputs: (manipulated variables (MV) = 2, measured disturbance (MD) = MV(3) = 1), and outputs (OV = 2) scale factors: MV (1) = 1, MV (2) = 0.9, MV (3) = 12 (the input MV span between the minimum and maximum value)
e. Specify constraints for MV, MD and MV Rate, MD Rate:

    MV (1). Min = 0.1, MV (1). Max = 1.1
    MV (1). RateMin = –0.1, MV (1). RateMax = 0.1
    MV (2). Min = 0.1, MV (2). Max = 1
    MV (2). RateMin = –0.1, MV (2). RateMax = 0.1
    MD (1). Min = 48, MV (3). Max = 60
    MD (1). RateMin = –0.1, MV (3). RateMax = 0.1

f. Specify constraint softening for MV and MV Rate:

    MV (1). RateMinECR = 100, MV (1). RateMaxECR = 1000
    MV (2). RateMinECR = 100, MV (2). RateMaxECR = 1000
    MV (3). RateMinECR = 100, MV (3). RateMaxECR = 1000

g. Specify constraints for OV:

OV (1). Min = 0, OV (1). Max = 12
OV (2). Min = 0, OV (2). Max = 50

h. Specify constraint softening for OV:

OV (1). MaxECR = 1000, OV (2). MaxECR = 1000

i. Specify overall adjustment factor applied to weights:
beta = 1.0833.

ii. Specify weights

Weights.MV = [0 0 0] *beta, Weights.MVRate = [0.09231 0.0923 0.0923]/beta
Weights.OV = [108.33 0.75831] *beta
Weights.ECR = 100000

k. Specify overall adjustment factor applied to estimation model gains:
alpha = 1.0965.

l. Adjust default output disturbance model gains:
setoutdist (mpc1, 'model', getoutdist(mpc1) *alpha).

m. Adjust default measurement noise model gains:
Model.Noise = Model.Noise/alpha.

n. Specify simulation options

options = mpcsimopt ()
options. MVSignal = mpc1_MVSignal
options. RefLookAhead = 'off'
options. MDLookAhead = 'off'
options. Constraints = 'on'
options. OpenLoop = 'off'

12.3.3 MIMO MPC MATLAB SIMULINK Simulation Results

The MATLAB simulation results of the MPC control performance in closed-loop for Centrifugal chiller plant based linear discrete-time MIMO ARMAX model are shown from Figs. 12.5, 12.6 and 12.7. In Fig. 12.5. is shown the evolution in closed-loop of the chilled water temperature in Evaporator subsystem, and in Fig. 12.6. is shown the controlled refrigerant liquid level in Condenser subsystem. Both figures reveal an excellent setpoint reference tracking performance, especially for chilled water temperature, fast transient, thus can be seen approximately 15 and 20 s for refrigerant liquid level in Condenser and chilled water temperature in Evaporator respectively. Also, is worth noticing a transient without oscillations, and only a small overshoot for chilled water temperature (8% from steady-state chilled water temperature).

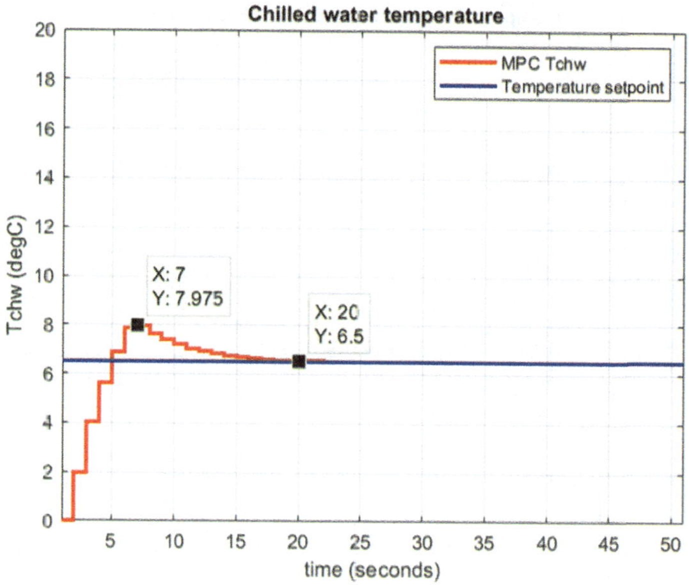

Fig. 12.5 MPC controller setpoint reference tracking performance for Evaporator chilled water temperature closed loop

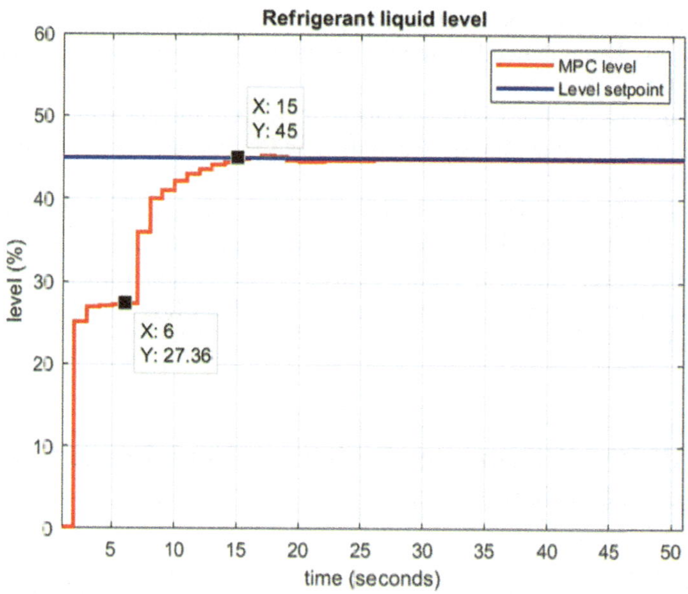

Fig. 12.6 MPC controller setpoint reference tracking performance for Condenser refrigerant liquid level closed loop

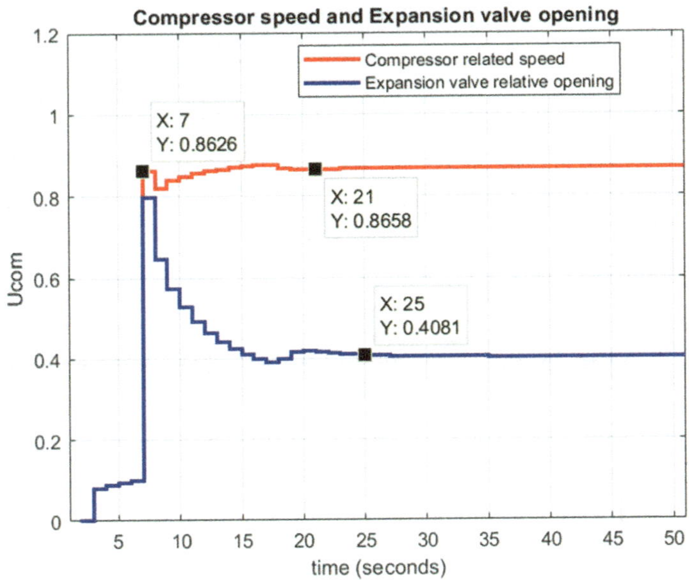

Fig. 12.7 The MPC relative compressor speed actuator effort

The smooth efforts of both actuators are shown in Fig. 12.7, while the expansion valve opening, while the chilled temperature load disturbance remains adjusted to $T_{rr} = 60[°C]$.

The robustness of MPC to changes in the reference setpoints is shown in Fig. 12.8 for chilled water temperature in Evaporator, and in Fig. 12.9 for refrigerant liquid level inCondenser.

In Fig. 12.10 can be seen smooth efforts from both actuators, while chilled water temperature load disturbance remains to the same previous adjusted value, $T_{rr} = 60$[degF].For lower values of T_{rr} we noticed a small setpoint tracking bias in the steady state refrigerant liquid level, thus it seems that MPC strategy behaves as a biased predictor.

12.4 MIMO MPC Strategy Design in a Nonminimal State Space Realization

The NMSS matrices F, G, H, D given in (12.9)–(12.15), according the development in [31], have the following elements:

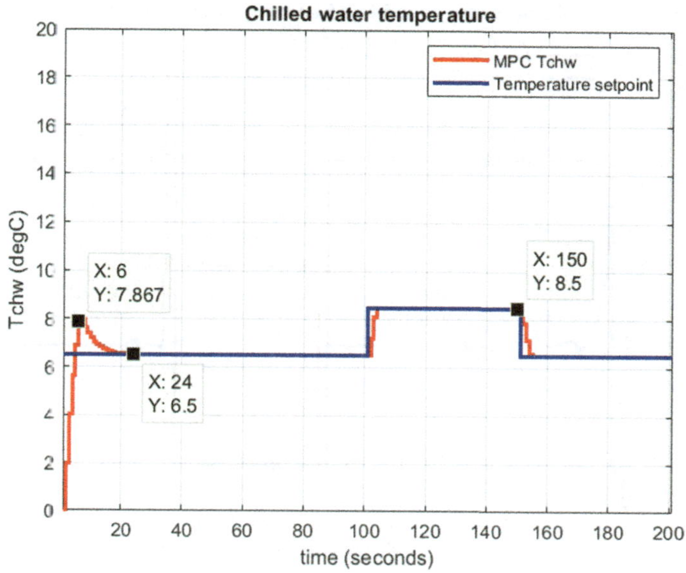

Fig. 12.8 The MPC controller tracking robustness performance to changes in reference setpoint of the Evaporator chilled water temperature closed loop

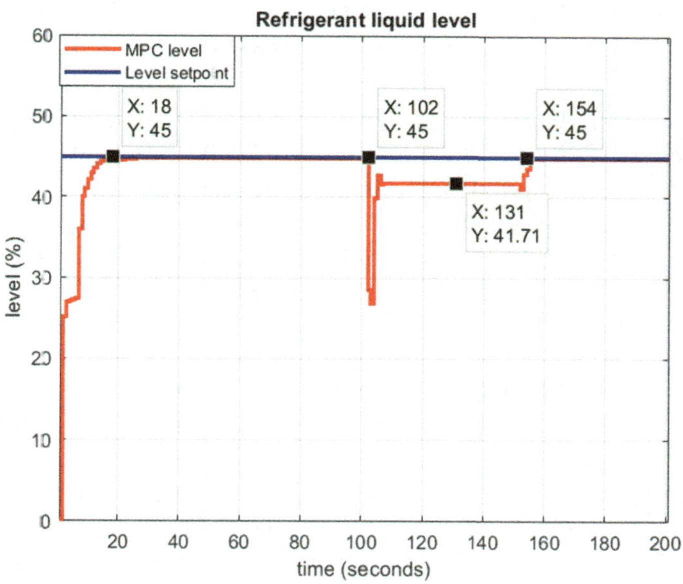

Fig. 12.9 MPC controller tracking performance robustness to changes in reference setpoint of the Condenser refrigerant liquid level closed loop

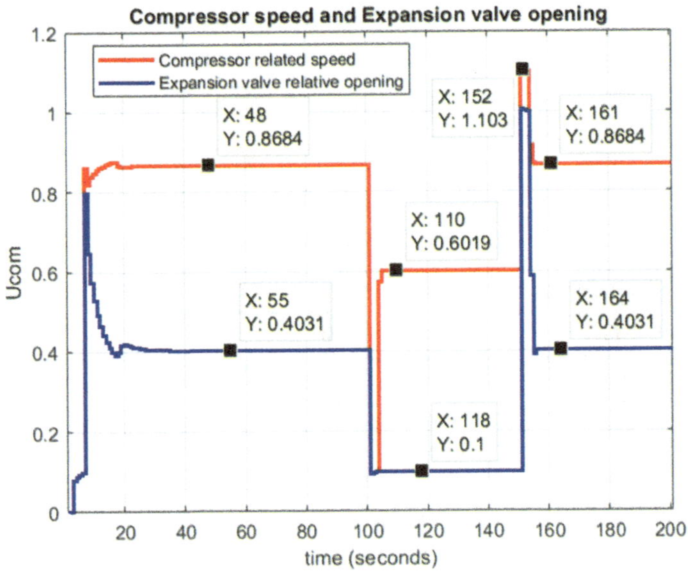

Fig. 12.10 The MPC relative compressor speed actuator effort to changes in the reference setpoint

$$F = \begin{bmatrix} 1.756 & 0 & -0.7611 & 0 & 0 & 0.0017 & 0 & 0 & 0 \\ 0 & 1.626 & 0 & -0.6884 & 0 & 2.071 & 0 & 0 & 0 \\ 1 & 0 & 0 & 0 & 0 & 0 & 0 & 0 & 0 \\ 0 & 1 & 0 & 0 & 0 & 0 & 0 & 0 & 0 \\ 0 & 0 & 0 & 0 & 0 & 0 & 0 & 0 & 0 \\ 0 & 0 & 0 & 0 & 0 & 0 & 0 & 0 & 0 \\ 0 & 0 & 0 & 0 & 0 & 0 & 0 & 0 & 0 \\ -1.756 & 0 & 0.7611 & 0 & 0 & -0.0017 & 0 & 1 & 0 \\ 0 & -1.626 & 0 & 0.6884 & 0 & -2.071 & 0 & 0 & 1 \end{bmatrix}$$

$$G = \begin{bmatrix} -0.0281 & -0.0081 & 0.0012 \\ 1.7760 & -2.9920 & 0.0305 \\ 0 & 0 & 0 \\ 0 & 0 & 0 \\ 1 & 0 & 0 \\ 0 & 1 & 0 \\ 0 & 0 & 1 \\ 0.0281 & 0.0081 & -0.0012 \\ -1.7760 & 2.9920 & -0.0305 \end{bmatrix}, D = \begin{bmatrix} 0 & 0 \\ 0 & 0 \\ 0 & 0 \\ 0 & 0 \\ 0 & 0 \\ 0 & 0 \\ 0 & 0 \\ 1 & 0 \\ 0 & 1 \end{bmatrix}, H = \begin{bmatrix} 1 & 0 \\ 0 & 1 \\ 0 & 0 \\ 0 & 0 \\ 0 & 0 \\ 0 & 0 \\ 0 & 0 \\ 1 & 0 \\ 0 & 1 \end{bmatrix}$$

Following the manual guidelines of SIMULINK Toolbox version R2018b provided in [34, 35], we start to build the MIMO MPC closed-loop strategy by opening the MPC Designer SIMULINK block, shown in Fig. 12.11, a very useful tool for parameters and weights tuning, as well as for closed-loop simulations.

The Design Review report shown in Fig. 12.12 provides a valuable information about the internal stability and the MPC success (PASS or FAIL) during closed-loop simulations using a linearized SIMULINK model setup shown in Fig. 12.13.

More details on MPC SIMULINK Designer can be found in [34] and [35]. The SIMULINK closed loop simulation results in terms of MIMO MPC tracking accuracy performance, fast transient speed, overshoot, and oscillations for both chiller outputs are shown in Figs. 12.14 and 12.15.

The actuators' efforts for the proposed MIMO MPC strategy in a NMSS design approach are shown in Fig. 12.16, for compressor speed, in Fig. 12.17 for expansion valve opening, and in Fig. 12.18 for chilled water temperature load disturbance.

The robustness of MIMO MPC strategy within a NMSS design approach to changes in the reference setpoints is shown in Figs. 12.19 and 12.20. The actuators efforts are shown in Fig. 12.21, for compressor speed, in Fig. 12.22 for expansion valve opening, and in Fig. 12.23 for chilled water temperature load disturbance.

Concluding, the MATLAB SIMULINK simulation results reveal a setpoint tracking MIMO MPC performance accuracy within a NMSS design approach greater than those obtained within a MSS design approach.

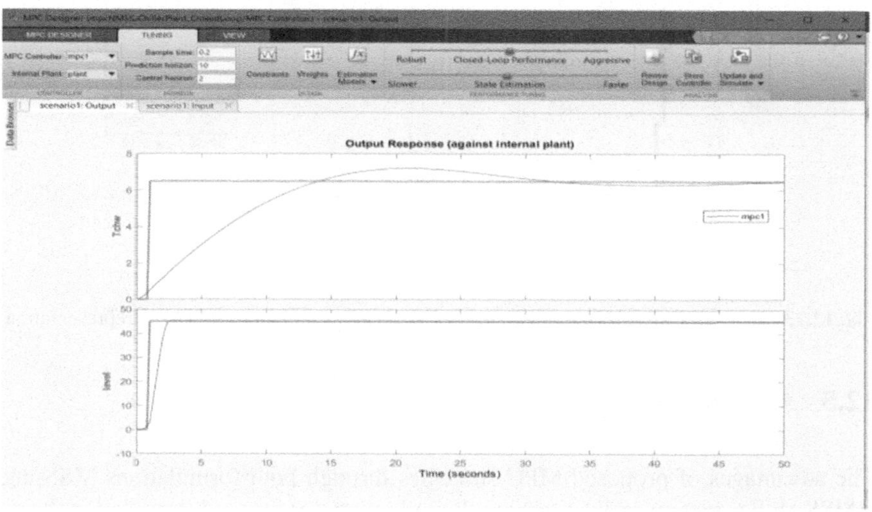

Fig. 12.11 MPC Designer SIMULINK block for control design and parameters tuning in NMSS representation

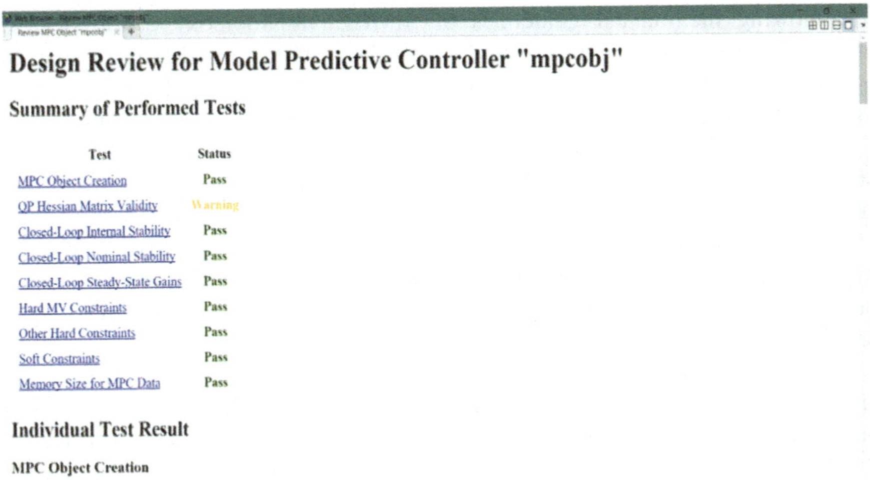

Fig. 12.12 SIMULINK block Design Review for MPC oriented object in NMSS representation

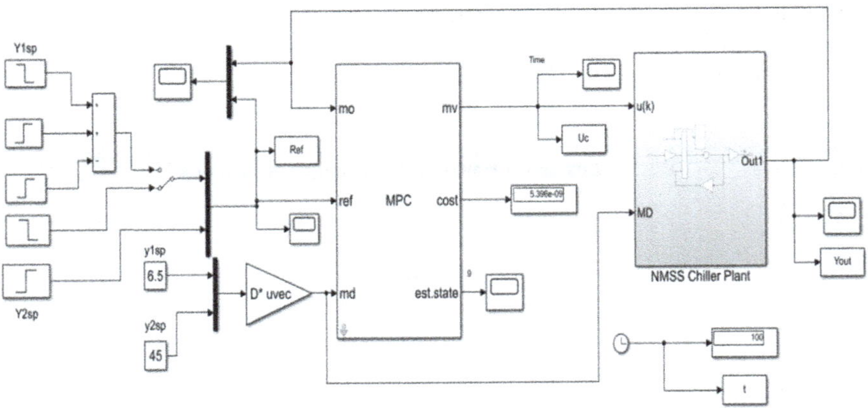

Fig. 12.13 MPC in SIMULINK model setup for MATLAB simulations in NMSS representation

12.5 Conclusions

The advantages of proposed MPC strategies through both formulations MSS and NMSS chiller system realizations are the following: (a) forward formulation, based on well-understood concepts, (b) explicitly handles constraints and use of the model, (c) well understood tuning parameters such as prediction and control horizon, sampling time, (d) more flexibility to set up the weights and weights rates for optimization problem formulation, (e) development time much shorter, (f) rejection of the input and output disturbances, as well as the measurement sensor noises, and

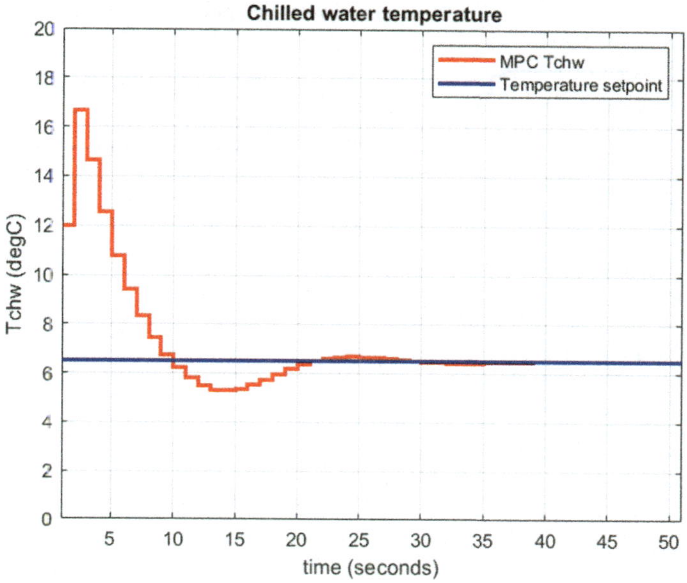

Fig. 12.14 MPC reference setpoint tracking performance for chilled water temperature in NMSS representation

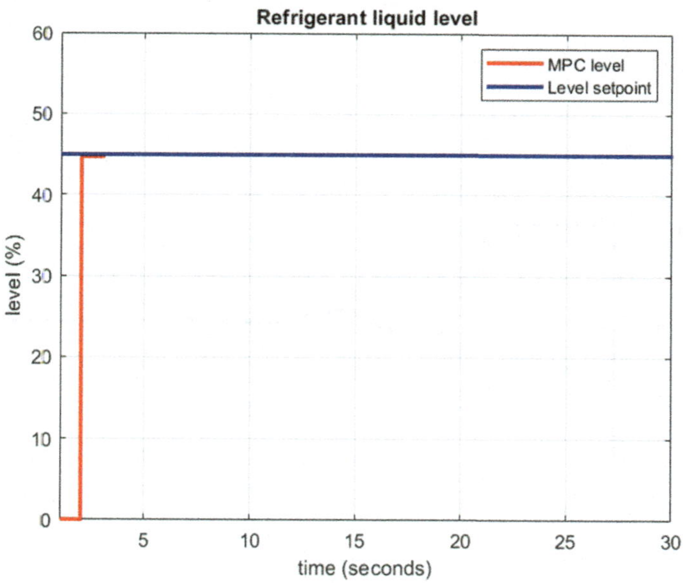

Fig. 12.15 MPC reference setpoint tracking performance for refrigerant liquid level in NMSS representation

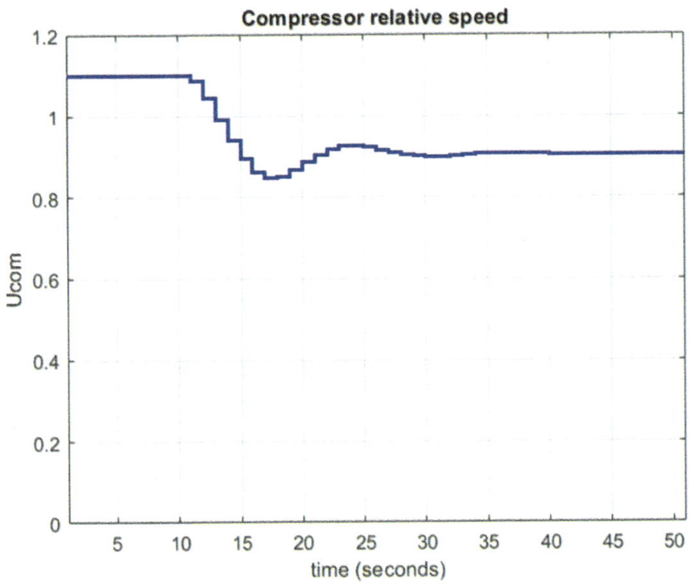

Fig. 12.16 The MPC relative compressor speed actuator effort in NMSS representation

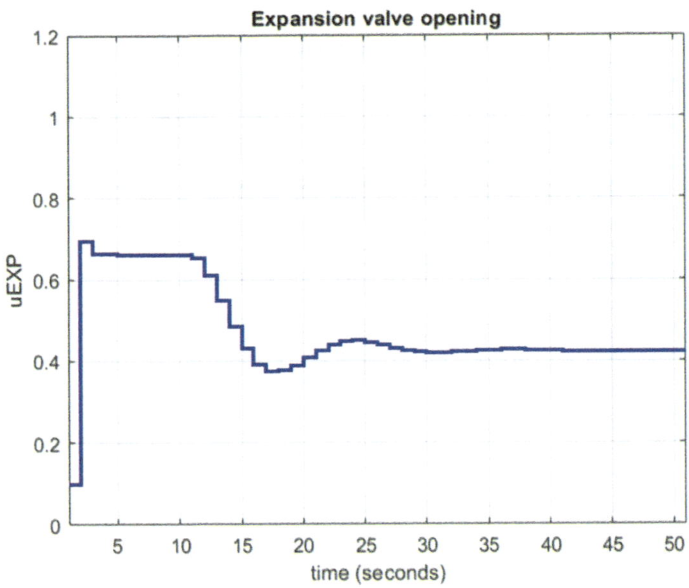

Fig. 12.17 The MPC relative expansion valve opening actuator effort in NMSS representation

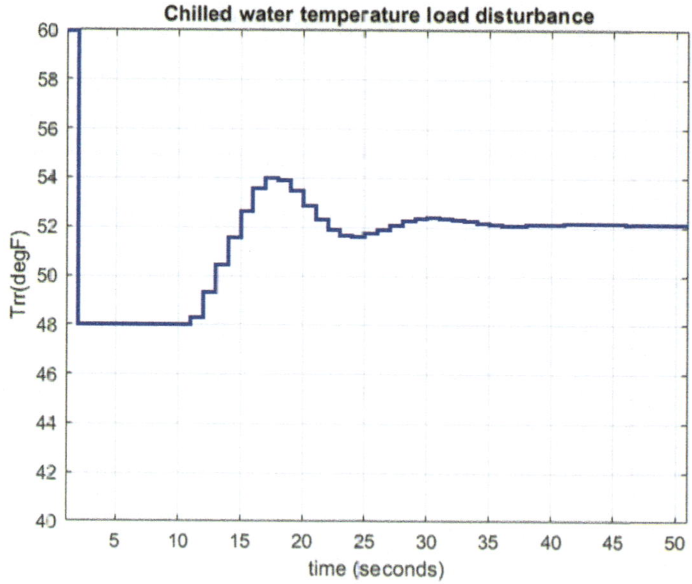

Fig. 12.18 The MPC evolution of chilled water load temperature in NMSS representation

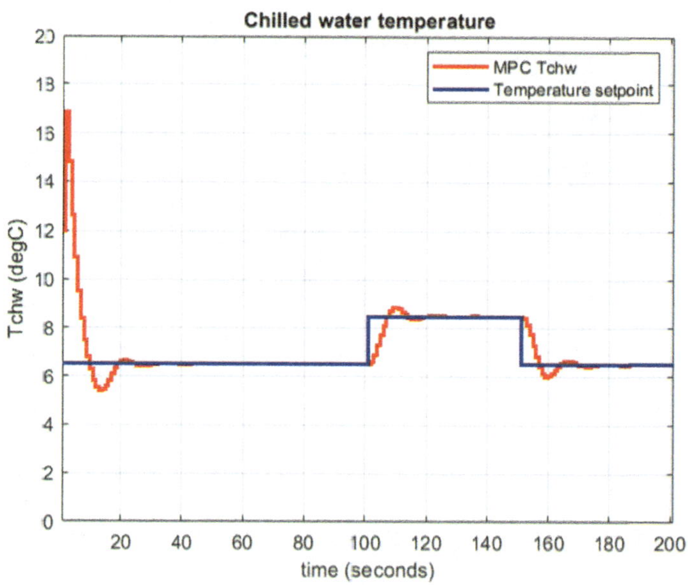

Fig. 12.19 The robustness of MPC to changes in reference setpoint for closed loop chilled water temperature in NMSS representation

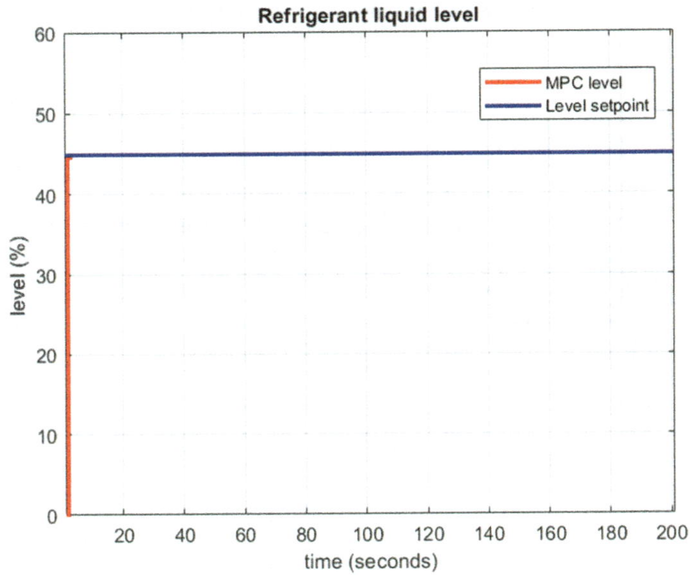

Fig. 12.20 The robustness of MPC to changes in reference setpoint for closed loop refrigerant liquid level in NMSS representation

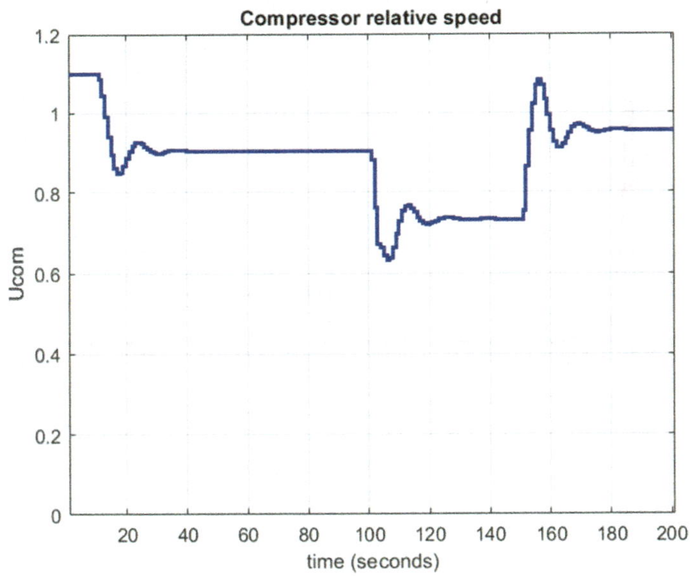

Fig. 12.21 The MPC relative compressor speed actuator effort in NMSS representation to changes in reference setpoint

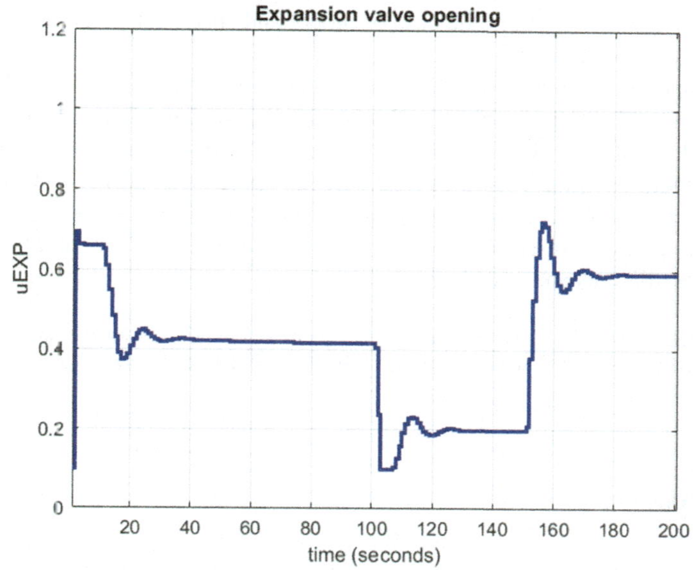

Fig. 12.22 The MPC relative expansion valve opening actuator effort in NMSS representation for changes in setpoint

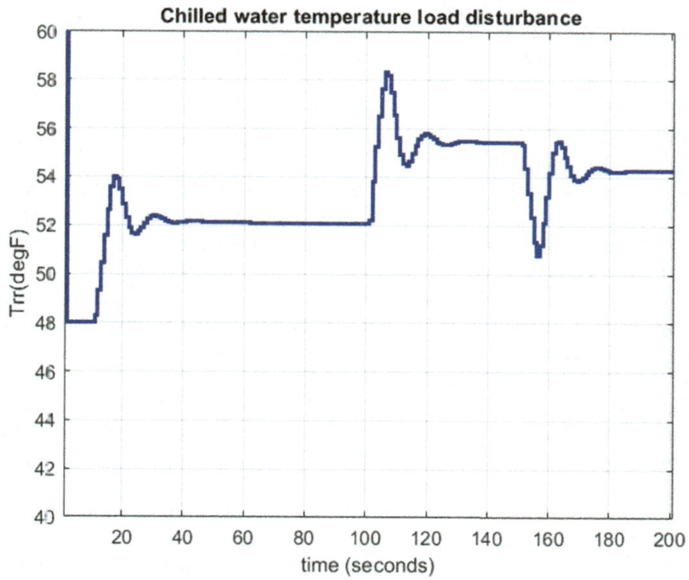

Fig. 12.23 The MPC evolution of chilled water load temperature in NMSS representation for changes in setpoints

(g) easier to maintain, i.e., in the case of model and specifications, changes do not require a complete redesign.

The proposed new MIMO MPC strategy can be applied for many control systems applications with high complexity, dimensionality, and nonlinearity. It can also be applied to MIMO skewed control systems, multiple control objectives, and low computational microprocessor load cost. The proposed MIMO MPC strategies proved effectiveness, excellent capability to outperform the competing advanced control strategies, setpoint tracking accuracy, fast transient speed, number of step response oscillations and overshoot.

Acknowledgements Research funding (discovery grant) for this project from the Natural Sciences and Engineering Research Council of Canada (NSERC) is gratefully acknowledged.

References

1. J.M. Gordon, K.C. Ng, H.T. Chua, Centrifugal chillers: Thermodynamic modelling and a diagnostic case study. Int. J. Ref. **18**(4), 253–257 (1995)
2. L. Pengfei, L. Yaoyu, J.E. Seem, Dynamic modelling and consistent initialization of system of differential-algebraic equations for centrifugal chillers, in *The Fourth National Conf. of IBPSA-USA* (New York, 2010), pp. 386–393
3. G.-Y. Jin, X.-D. Ding, P.-Y. Tan et al., A hybrid water-cooled centrifugal chiller model, in *6th IEEE Conference on Industrial Electronics and Applications* (IEEEXplore Digital Library, Beijing, China, 2011), pp. 2298–2303
4. S. Huang, W. Zuo, M.D. Sohn, A new method for the optimal chiller sequencing control, proceedings of BS2015, in *14th Conference of International Building Performance Simulation Association* (Hyderabad, India, 2015), pp. 316–323
5. L. Pengfei, L. Yaoyu, J.E. Seem et al., Recent advances in dynamic modeling of HVAC equipment. Part 2: Modelica—based modellin. J. HVAC&R Res. **20**(1), 150–16 (2014)
6. H. Wang, A steady-state empirical model for evaluating energy efficient performance of centrifugal water chillers. Energy Build, Elsevier **154**, 415–419 (2017)
7. N. Tudoroiu, M. Zaheeruddin, Li Songchun, R-E Tudoroiu: Design and implementation of closed-loop pi control strategies in real-time MATLAB simulation environment for nonlinear and linear ARMAX models of HVAC centrifugal chiller control systems', advances in science. Technol. Eng. Syst. J. (ASTESJ) **3**(2), 283–308 (2018)
8. R.-E. Tudoroiu, M. Zaheeruddin, N. Tudoroiu, D.D. Burdescu, MATLAB Implementation of an Adaptive Neuro-Fuzzy Modeling Approach applied on Nonlinear Dynamic Systems – a Case Study. Proc. Federated Conf. Comput. Sci. Inf. Syst. ACSIS **15**, 577–583 (2018)
9. I.D. Landau, R. Lozano, M. M'Saad, A. Karimi, *Discrete-Time System Models for Control -Chapter 2* (Adaptive Control, Communications and Control Engineering (Springer, London Limited, 2011), pp. 35–53
10. A.A. Nada, E.M. Shaban, The development of proportional-integral-plus control using field programmable gate array technology applied to mechatronics system. American J. Res. Com. **2**(4), 14–27 (2014)
11. E.M. Shaban, A.A. Nada, Proportional integral derivative versus proportional integral plus control applied to mobile robotic system. J. Am. Sci. **9**(12), pp. 583–591, (2013).
12. C.J. Taylor, A. Chotai, P.C. Young, Proportional-integral-plus (PIP) control of time delay systems. Proc. Inst. Mech. Eng. **212**(1), 37–47 (1998)

13. C.J. Taylor, E.M. Shaban, Multivariable Proportional-Integral-Plus (PIP) control of the ALSTOM nonlinear gasifier simulation. IEE Proc. Control Theory Appl. **153**(3), 277–285 (2006)
14. P.C. Young, A.P. McCabe, A. Chotai, State-dependent parameter nonlinear systems: Identification. Estimation and Control, IFAC, 15th Triennial World Congress (Barcelona, Spain, 2002)
15. V. Exadaktylos, Model Predictive Control Structures in the Non-Minimal State Space. Ph.D. thesis. Lancaster University, UK (2007)
16. L. Lennart, *System Identification Toolbox Getting Started Guide* (MathWorks, Inc., 2018), pp. 1–220. https://www.mathworks.com/help/pdf_doc/ident/ident_gs.pdf, accessed 10 November 2018
17. B. Messner, D. Tilbury, *Control Tutorials for MATLAB* (Carnegie Mellon Lab, University of Michigan). http://ctms.engin.umich.edu.
18. MathWorks MATLAB Version R2018b—Model Predictive Control Toolbox On-lineDocumentation: https://www.mathworks.com/help/mpc/index.html.
19. A. Bemporad, M. Morari, N.L. Ricker, *Model Predictive Control Toolbox User's Guide* (The MathWorks, Inc., 2018), pp. 1–942. Available also on-line at: https://www.mathworks.com/help/pdf_doc/mpc/mpc_ug.pdf
20. T. Takahama, D. Akasaka, *Society of Automotive Engineers of Japan: Model Predictive Control Approach to Design Practical Adaptive Cruise Control for Traffic Jam* (2017), pp.1–6. available also on-line at: https://www.mathworks.com/content/dam/mathworks/tag-team/Objects/m/FASTzero17_MPC4ACC_20170719.pdf, accessed 10 November 2018
21. Y.J. Kim, C.S. Park, nonlinear predictive of chiller system using gaussian process model, in *IBPSA Asia Conference (ASimo2014)* (Nagoya, Japan, 2014), pp. 594–601
22. O. Mikuláš, A Framework for Nonlinear Model Predictive Control, Master thesis, Czeck Technical University, Prague (2016)
23. S. Gros, M. Zanon, R. Quirynen et al., From linear to nonlinear MPCbridging the gap via the real-time iteration. Int. J. Control 1–19 (2016)
24. M. Zanon, A. Boccia, V.G.S Palma et al., Direct Optimal control and model predictive control, in *Lecture Notes in MathematOptimal Control - Novel Directions and Applications*, vol. 2180, eds. by D. Tonon, M. Soledad Aronna, D. Kalise (Springer International Publishing, 2017), pp. 263–382

The manufacturer's authorised representative in the EU is Springer Nature Customer Service Centre GmbH, Europaplatz 3, 69115 Heidelberg, Germany. If you have any concerns regarding our products, please contact ProductSafety@springernature.com

Printed and bound by CPI Group (UK) Ltd, Croydon, CR0 4YY
30/09/2026
02225293-0004